A VISUAL ATLAS FOR

Anatomy and Physiology

Featuring art from

Kenneth S. Saladin
Anatomy and Physiology, 3rd Edition

McGraw Hill

Boston   Burr Ridge, IL   Dubuque, IA   Madison, WI   New York   San Francisco   St. Louis
Bangkok   Bogotá   Caracas   Kuala Lumpur   Lisbon   London   Madrid   Mexico City
Milan   Montreal   New Delhi   Santiago   Seoul   Singapore   Sydney   Taipei   Toronto

The McGraw·Hill Companies

A VISUAL ATLAS FOR ANATOMY AND PHYSIOLOGY

Published by McGraw-Hill Higher Education, an imprint of The McGraw-Hill Companies, Inc., 1221 Avenue of the Americas, New York, NY 10020. Copyright © The McGraw-Hill Companies, Inc., 2004. All rights reserved.

All photos are © McGraw-Hill Higher Education, Inc./Rebecca Gray, photographer/Don Kincaid, dissections; except
Figure 9.26: From Vidic/Saurez, *Photographic Atlas of the Human Body.* ©1984 Mobsy-Year Book, Inc., St. Louis, MO; Figure 19.11a: © Ed Reschke

This book is printed on acid-free paper.

7 8 9 0 VHM VHM 0 9 8 7 6

ISBN-13: 978-0-07-294394-8
ISBN-10: 0-07-294394-7
www.mhhe.com

Contents

Unit 4 The Muscular System   61

Unit 5 The Spinal Cord and Spinal Nerves   102

Unit 6 The Brain and Cranial Nerves   115

Unit 7 The Sense Organs   139

Unit 8 The Heart   146

Unit 9 The Respiratory System   156

Preface

This collection of images was assembled to provide students with a comprehensive resource for studying anatomical structures and a convenient place to write notes during lecture or lab. The drawings and photographs presented in this atlas have been enlarged to the maximum size the page dimensions will allow, providing large, clear images with enhanced detail.

This atlas features full coverage of the gross anatomy of the skeletal system, the muscular system, and major joints. Also included are selected illustrations and photographs of key anatomical structures from the nervous system, as well as images depicting the anatomy of key organs and sensory structures.

In creating the *Visual Atlas for Anatomy and Physiology*, McGraw-Hill aims to meet the needs of the many different types of students enrolled in anatomy and physiology courses by providing tools that complement various learning styles. If you have any suggestions for how this product can be improved in future editions, please send your comments to the address below.

Martin J. Lange
Publisher, Life Sciences
McGraw-Hill Higher Education
2460 Kerper Boulevard
Dubuque, IA 52001

Unit 1: The Axial Skeleton

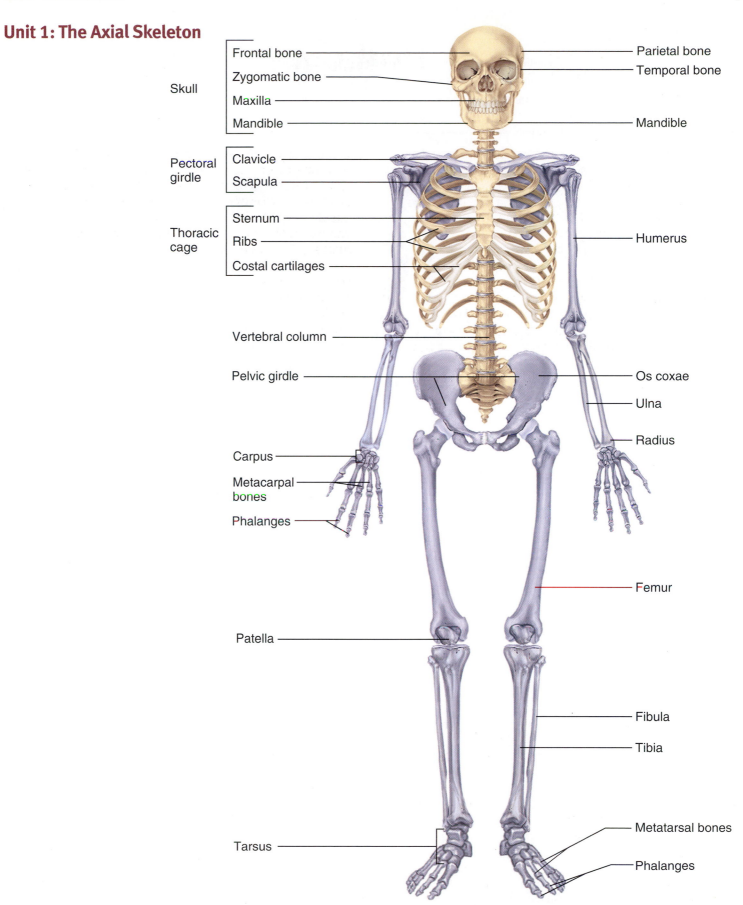

Frontal bone
Zygomatic bone
Maxilla
Mandible

Skull

Parietal bone
Temporal bone

Mandible

Pectoral girdle
Clavicle
Scapula

Thoracic cage
Sternum
Ribs
Costal cartilages

Humerus

Vertebral column

Pelvic girdle

Os coxae

Ulna

Radius

Carpus

Metacarpal bones

Phalanges

Femur

Patella

Fibula

Tibia

Metatarsal bones

Tarsus

Phalanges

Figure 8.1a   The Adult Skeleton

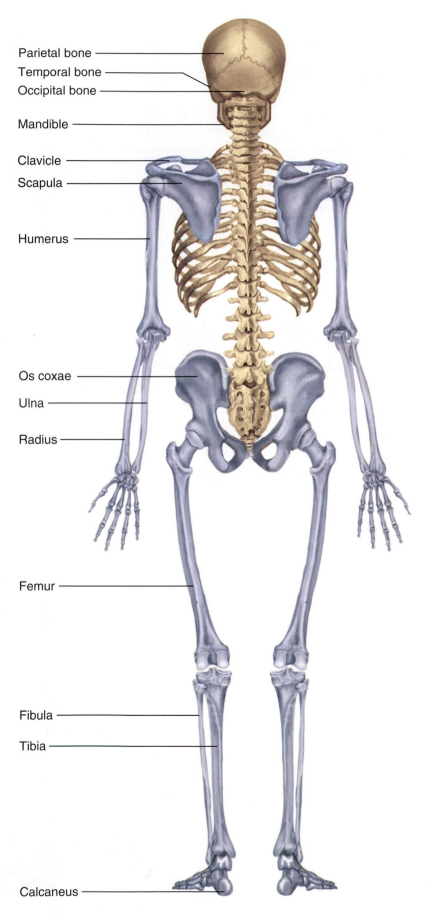

Parietal bone
Temporal bone
Occipital bone
Mandible
Clavicle
Scapula
Humerus
Os coxae
Ulna
Radius
Femur
Fibula
Tibia
Calcaneus

Figure 8.1b   The Adult Skeleton

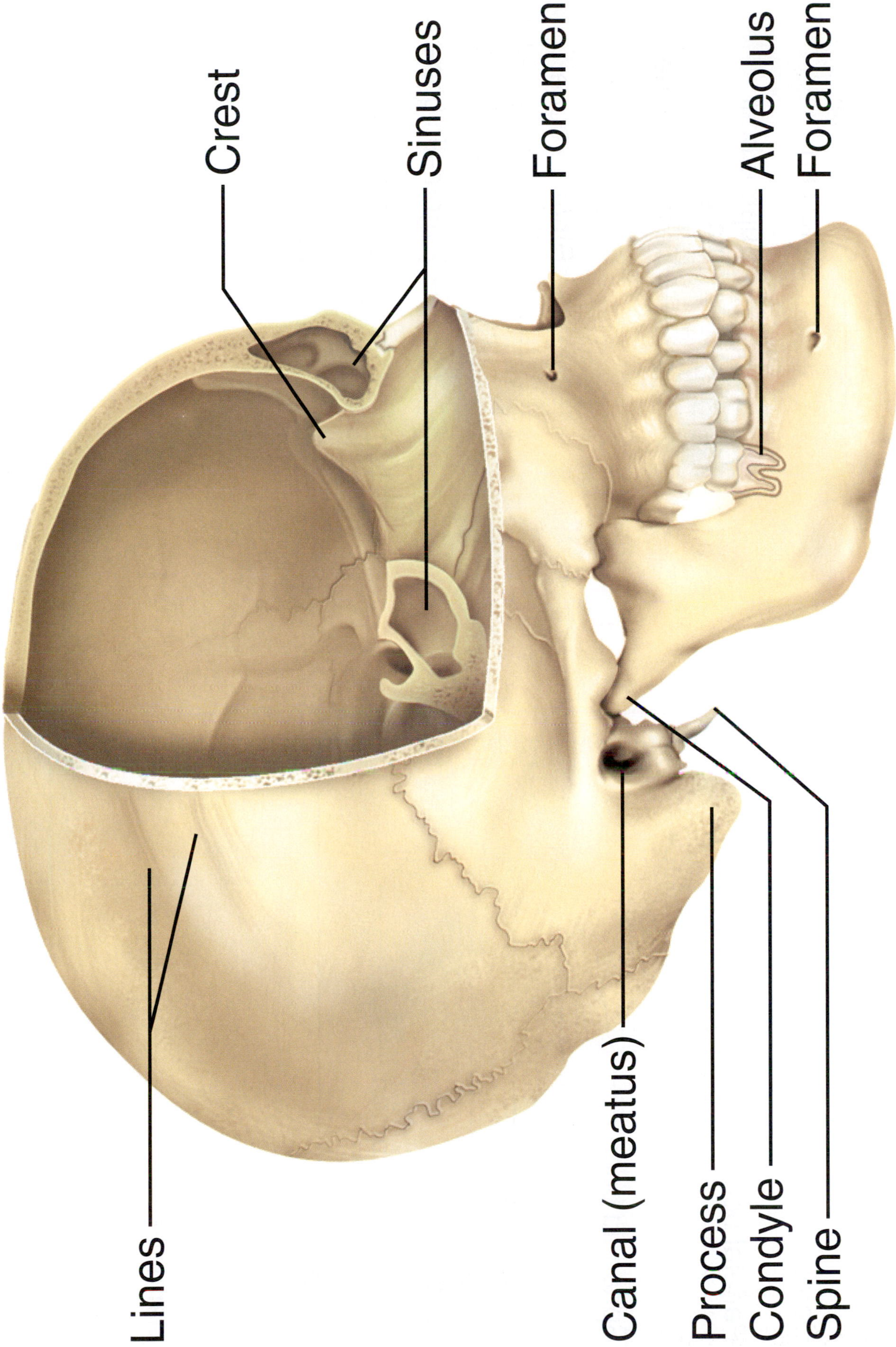

Crest

Sinuses

Foramen

Alveolus

Foramen

Lines

Canal (meatus)

Process

Condyle

Spine

Figure 8.2a Surface Features of Bones

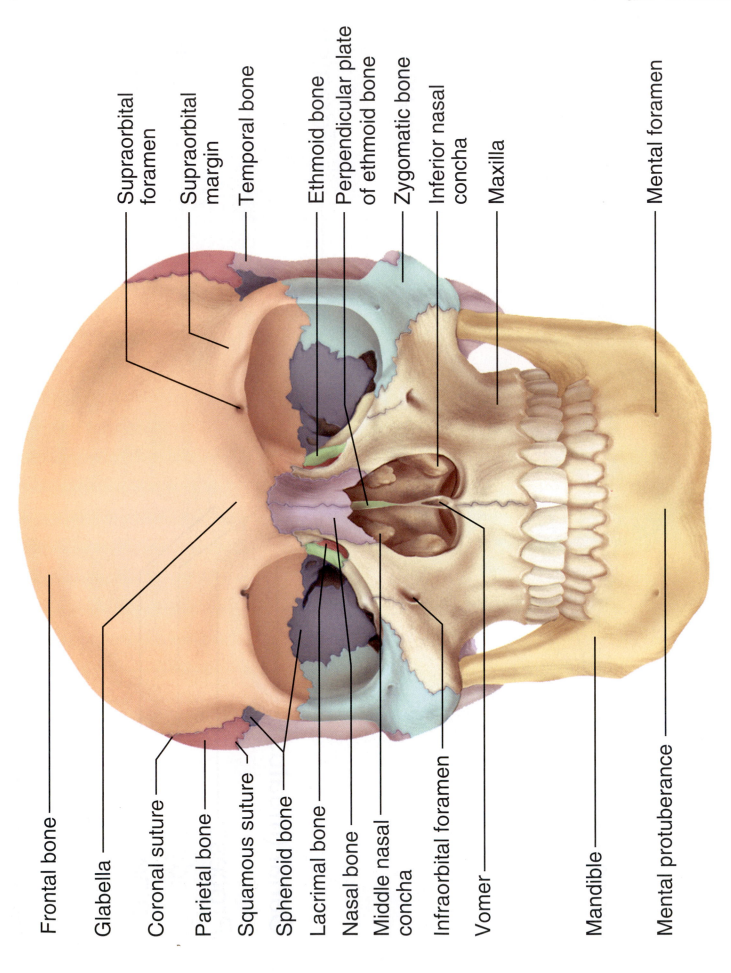

Supraorbital foramen

Supraorbital margin

Temporal bone

Ethmoid bone

Perpendicular plate of ethmoid bone

Zygomatic bone

Inferior nasal concha

Maxilla

Mental foramen

Frontal bone

Glabella

Coronal suture

Parietal bone

Squamous suture

Sphenoid bone

Lacrimal bone

Nasal bone

Middle nasal concha

Infraorbital foramen

Vomer

Mandible

Mental protuberance

Figure 8.3   The Skull, Anterior View

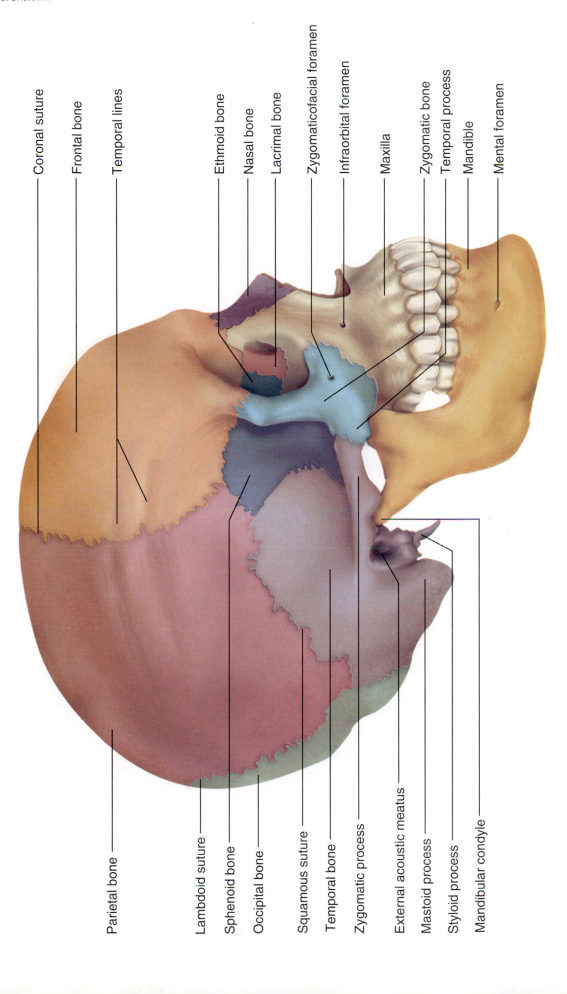

Coronal suture

Frontal bone

Temporal lines

Ethmoid bone

Nasal bone

Lacrimal bone

Zygomaticofacial foramen

Infraorbital foramen

Maxilla

Zygomatic bone

Temporal process

Mandible

Mental foramen

Parietal bone

Lambdoid suture

Sphenoid bone

Occipital bone

Squamous suture

Temporal bone

Zygomatic process

External acoustic meatus

Mastoid process

Styloid process

Mandibular condyle

Figure 8.4a The Skull

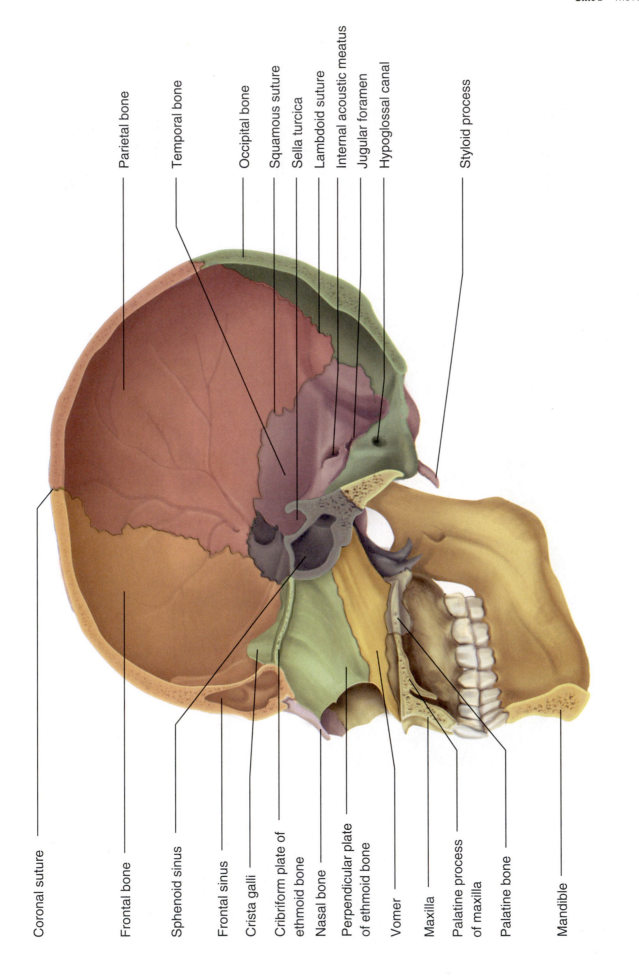

Parietal bone

Temporal bone

Occipital bone

Squamous suture

Sella turcica

Lambdoid suture

Internal acoustic meatus

Jugular foramen

Hypoglossal canal

Styloid process

Coronal suture

Frontal bone

Sphenoid sinus

Frontal sinus

Crista galli

Cribriform plate of ethmoid bone

Nasal bone

Perpendicular plate of ethmoid bone

Vomer

Maxilla

Palatine process of maxilla

Palatine bone

Mandible

Figure 8.4b The Skull

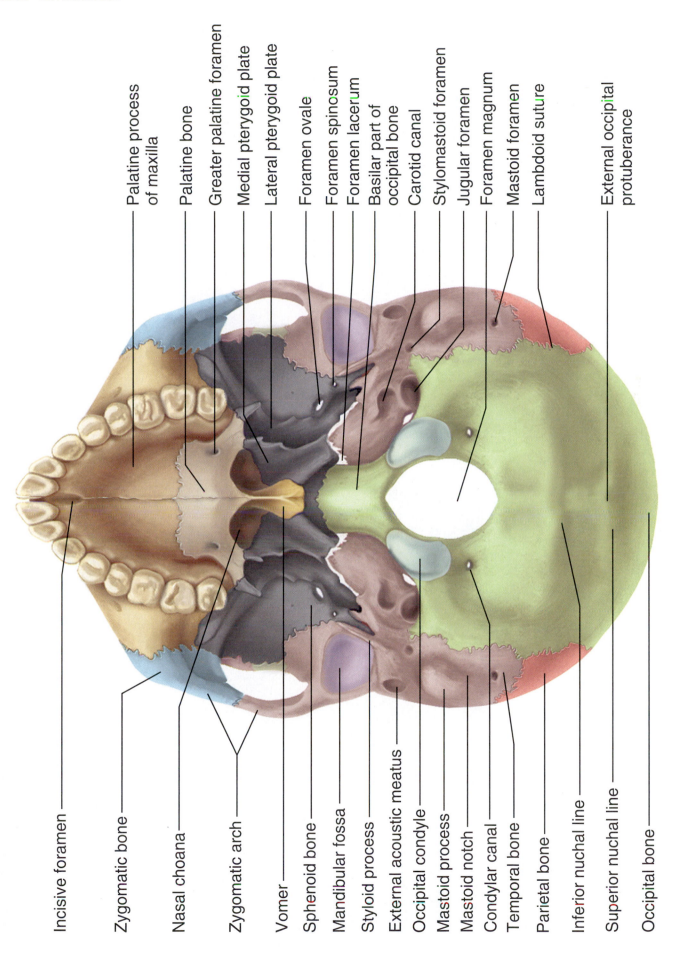

Palatine process of maxilla
Palatine bone
Greater palatine foramen
Medial pterygoid plate
Lateral pterygoid plate
Foramen ovale
Foramen spinosum
Foramen lacerum
Basilar part of occipital bone
Carotid canal
Stylomastoid foramen
Jugular foramen
Foramen magnum
Mastoid foramen
Lambdoid suture
External occipital protuberance

Incisive foramen
Zygomatic bone
Nasal choana
Zygomatic arch
Vomer
Sphenoid bone
Mandibular fossa
Styloid process
External acoustic meatus
Occipital condyle
Mastoid process
Mastoid notch
Condylar canal
Temporal bone
Parietal bone
Inferior nuchal line
Superior nuchal line
Occipital bone

Figure 8.5a Base of the Skull

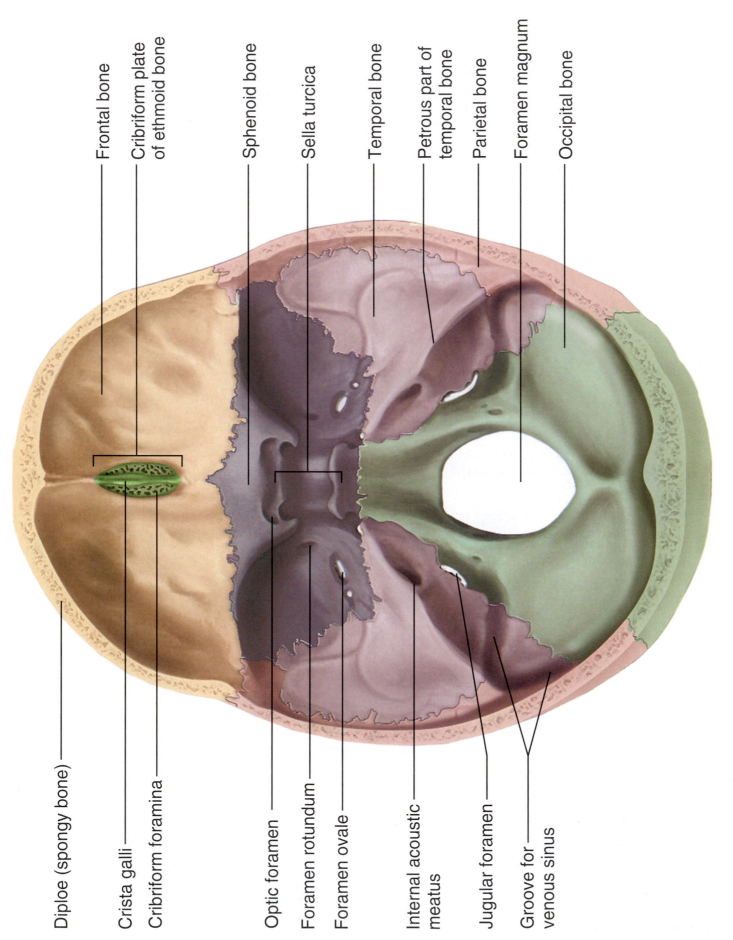

Diploe (spongy bone)

Crista galli

Cribriform foramina

Frontal bone

Cribriform plate
of ethmoid bone

Sphenoid bone

Sella turcica

Temporal bone

Petrous part of
temporal bone

Parietal bone

Foramen magnum

Occipital bone

Optic foramen

Foramen rotundum

Foramen ovale

Internal acoustic
meatus

Jugular foramen

Groove for
venous sinus

Figure 8.5b Base of the Skull

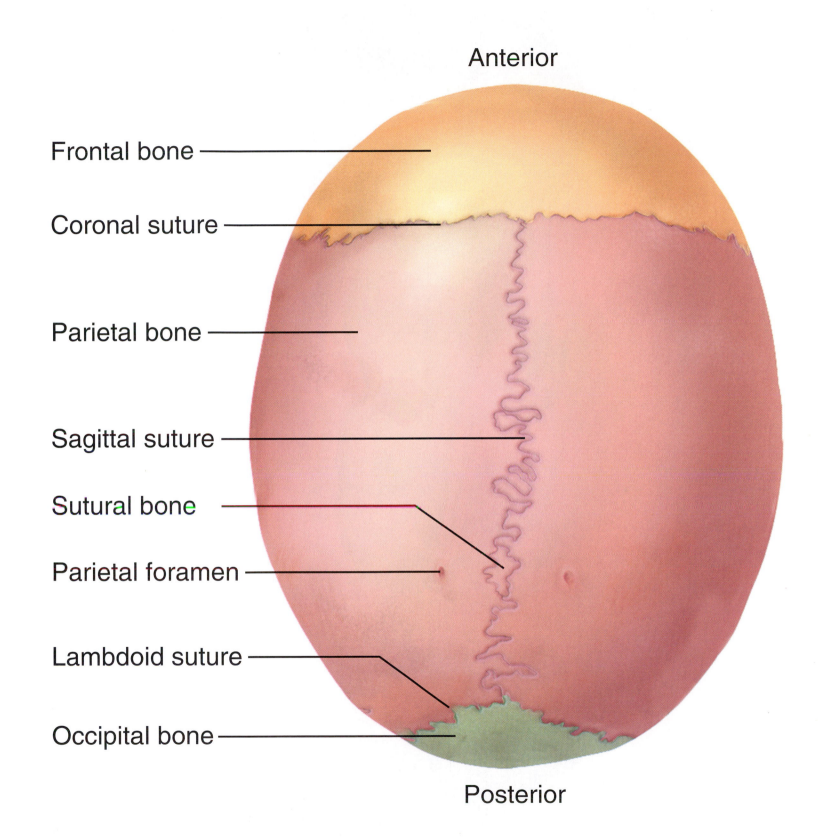

Anterior

Frontal bone

Coronal suture

Parietal bone

Sagittal suture

Sutural bone

Parietal foramen

Lambdoid suture

Occipital bone

Posterior

Figure 8.6 **The Calvaria, Superior View**

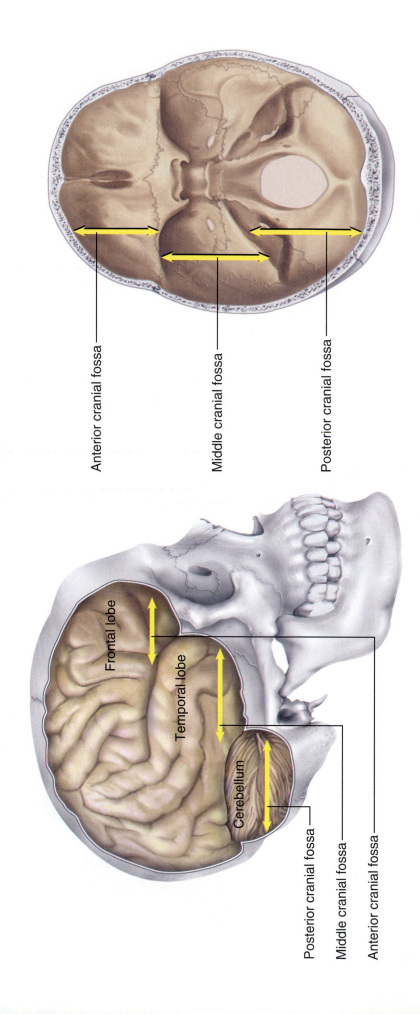

Anterior cranial fossa

Middle cranial fossa

Posterior cranial fossa

Frontal lobe

Temporal lobe

Cerebellum

Posterior cranial fossa

Middle cranial fossa

Anterior cranial fossa

Figure 8.9   Cranial Fossae

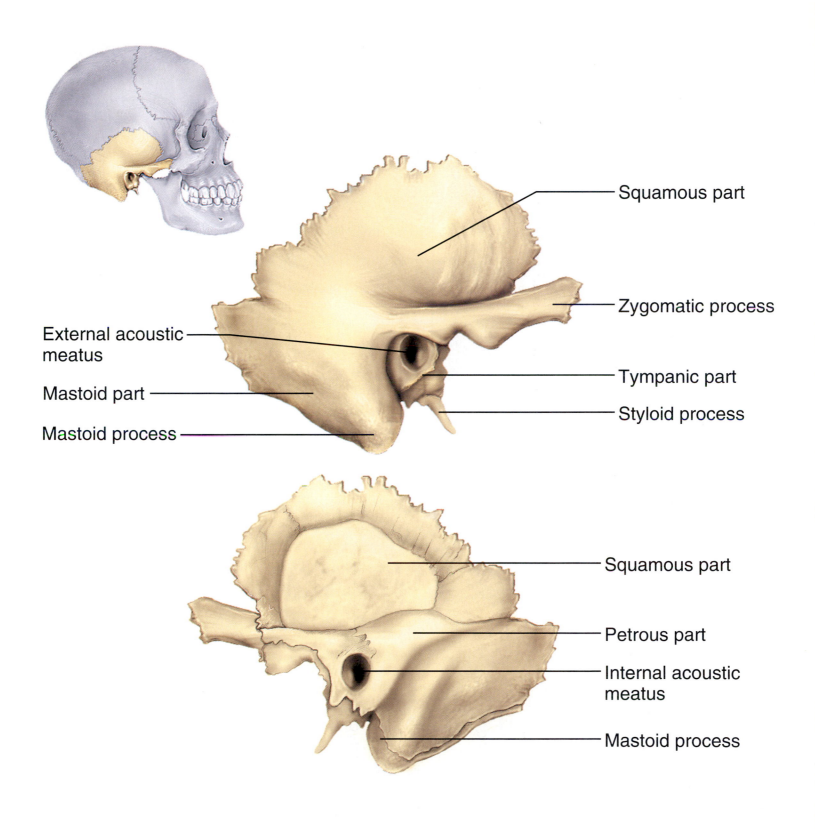

Squamous part

Zygomatic process

External acoustic meatus

Tympanic part

Mastoid part

Styloid process

Mastoid process

Squamous part

Petrous part

Internal acoustic meatus

Mastoid process

Figure 8.10   The Right Temporal Bone

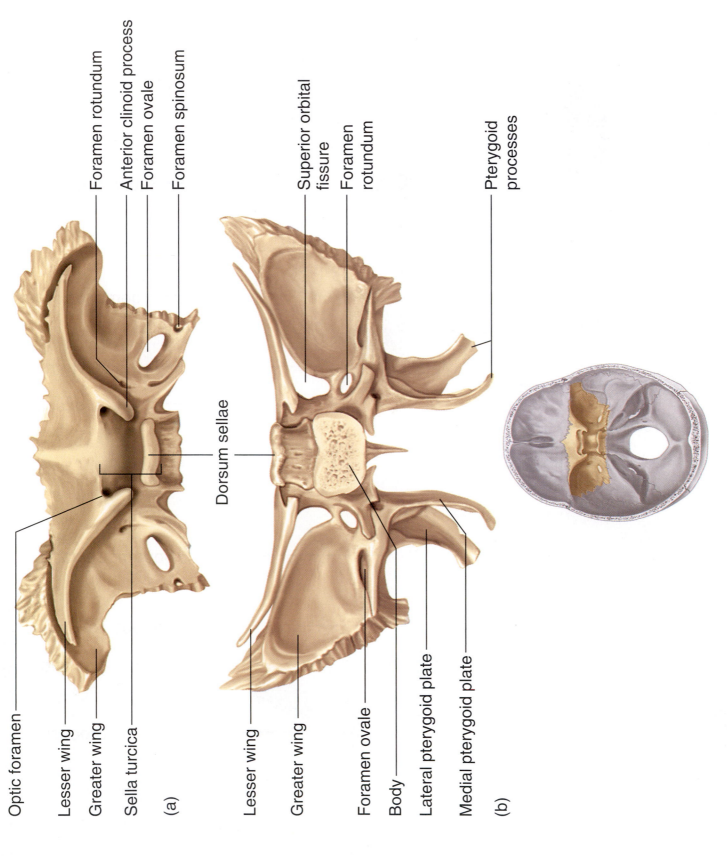

Optic foramen

Lesser wing

Greater wing

Sella turcica

(a)

Foramen rotundum

Anterior clinoid process

Foramen ovale

Foramen spinosum

Dorsum sellae

Lesser wing

Greater wing

Foramen ovale

Body

Lateral pterygoid plate

Medial pterygoid plate

(b)

Superior orbital fissure

Foramen rotundum

Pterygoid processes

Figure 8.11 The Sphenoid Bone

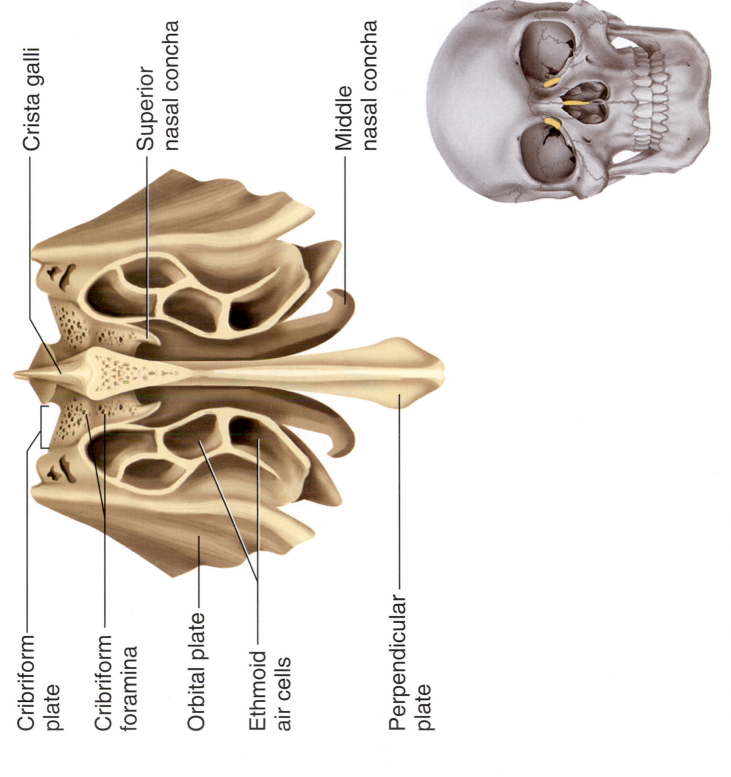

Crista galli

Superior nasal concha

Middle nasal concha

Cribriform plate

Cribriform foramina

Orbital plate

Ethmoid air cells

Perpendicular plate

Figure 8.12 The Ethmoid Bone, Anterior View

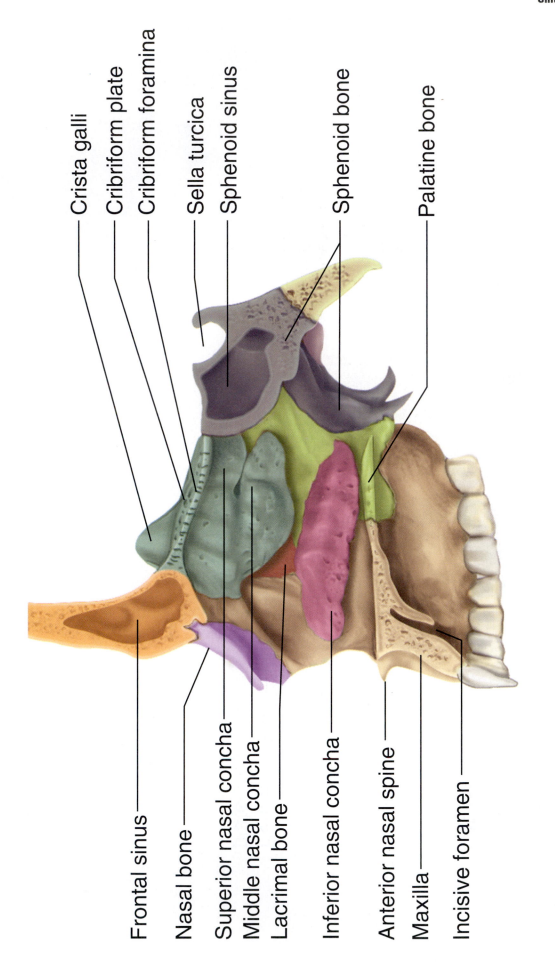

Crista galli
Cribriform plate
Cribriform foramina
Sella turcica
Sphenoid sinus
Sphenoid bone
Palatine bone

Frontal sinus
Nasal bone
Superior nasal concha
Middle nasal concha
Lacrimal bone
Inferior nasal concha
Anterior nasal spine
Maxilla
Incisive foramen

Figure 8.13   The Right Nasal Cavity, Sagittal Section

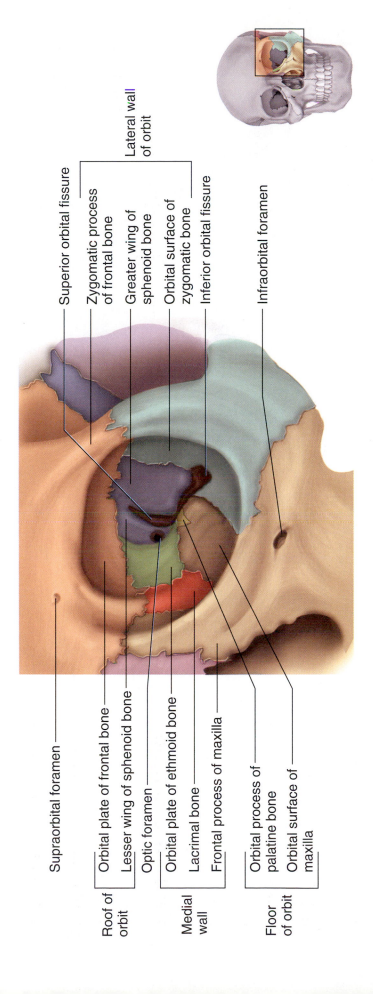

Superior orbital fissure

Zygomatic process of frontal bone

Greater wing of sphenoid bone

Orbital surface of zygomatic bone

Inferior orbital fissure

Lateral wall of orbit

Infraorbital foramen

Supraorbital foramen

Orbital plate of frontal bone

Lesser wing of sphenoid bone

Optic foramen

Roof of orbit

Orbital plate of ethmoid bone

Lacrimal bone

Frontal process of maxilla

Medial wall

Orbital process of palatine bone

Orbital surface of maxilla

Floor of orbit

Figure 8.14 The Left Orbit, Anterior View

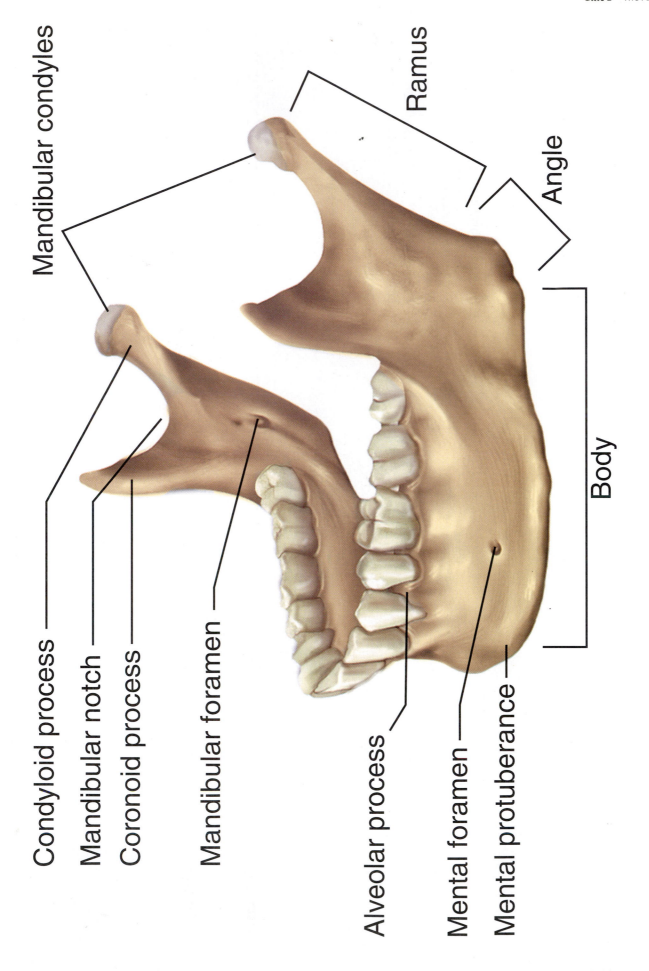

Mandibular condyles

Ramus

Angle

Body

Condyloid process

Mandibular notch

Coronoid process

Mandibular foramen

Alveolar process

Mental foramen

Mental protuberance

Figure 8.15   The Mandible

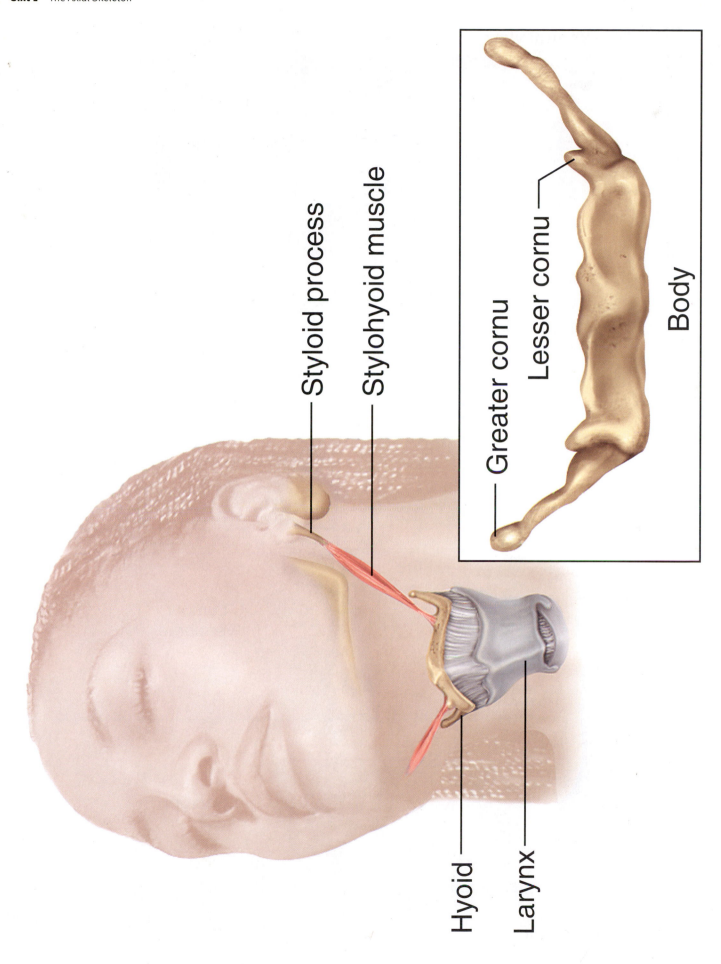

Styloid process

Stylohyoid muscle

Greater cornu

Lesser cornu

Body

Hyoid

Larynx

Figure 8.16   The Hyoid Bone

Ventral view   **Dorsal view**

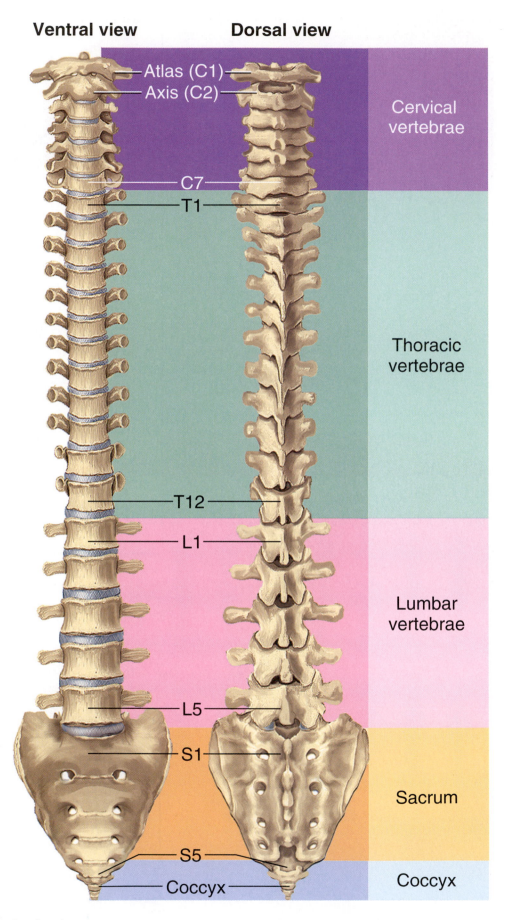

Atlas (C1)
Axis (C2)
Cervical vertebrae

C7
T1
Thoracic vertebrae

T12
L1
Lumbar vertebrae

L5
S1
Sacrum

S5
Coccyx
Coccyx

Figure 8.18   Vertebral column

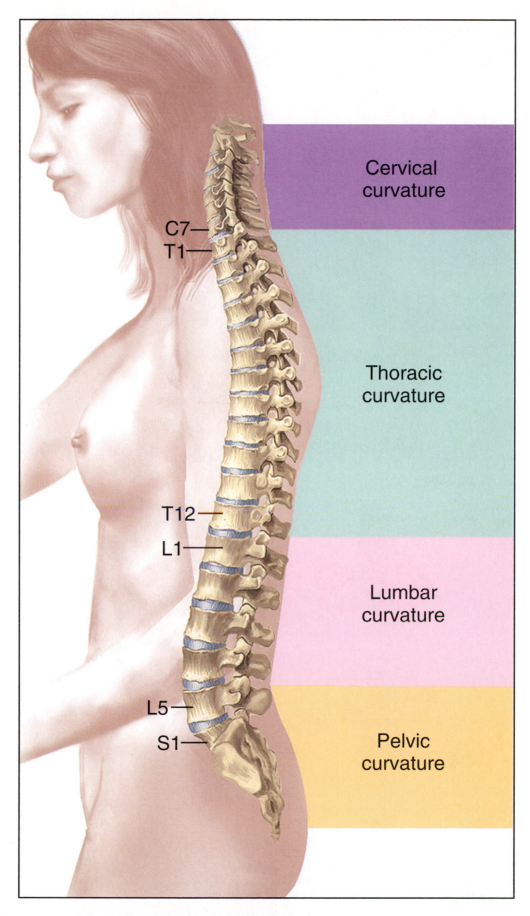

C7

T1

T12

L1

L5

S1

Cervical
curvature

Thoracic
curvature

Lumbar
curvature

Pelvic
curvature

Figure 8.19   Curvature of the Adult Vertebral Column

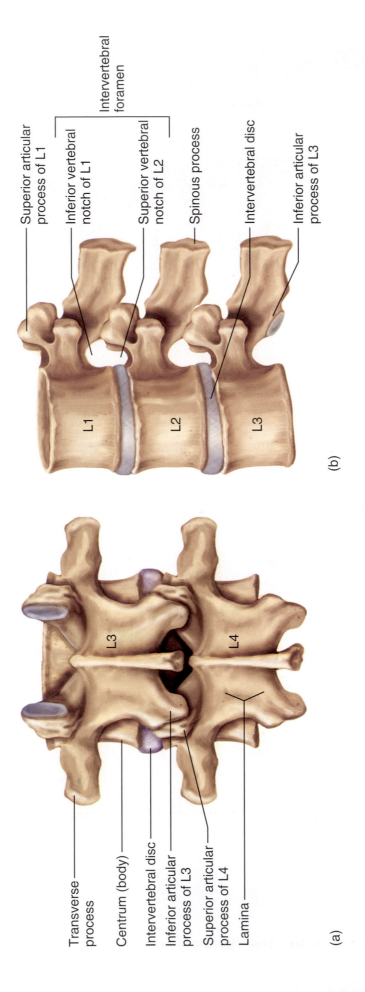

Superior articular process of L1

Inferior vertebral notch of L1

Intervertebral foramen

Superior vertebral notch of L2

Spinous process

Intervertebral disc

Inferior articular process of L3

L1

L2

L3

(b)

Transverse process

Centrum (body)

Intervertebral disc

Inferior articular process of L3

Superior articular process of L4

Lamina

L3

L4

(a)

Figure 8.23   Articulated Vertebrae

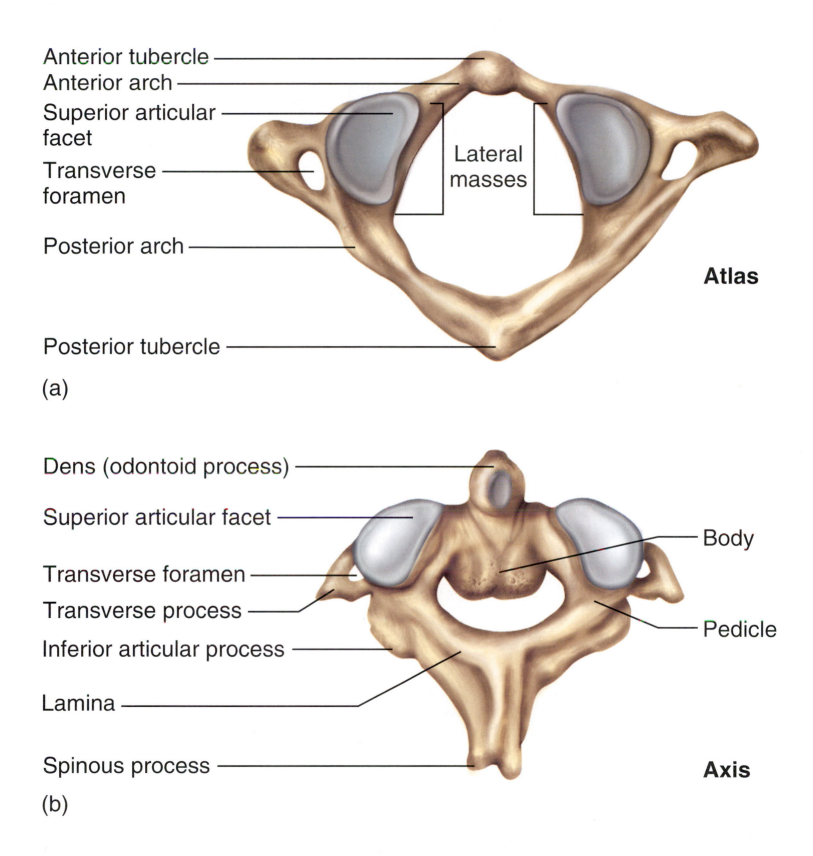

Anterior tubercle

Anterior arch

Superior articular facet

Transverse foramen

Lateral masses

Posterior arch

Atlas

Posterior tubercle

(a)

Dens (odontoid process)

Superior articular facet

Transverse foramen

Transverse process

Inferior articular process

Lamina

Body

Pedicle

Spinous process

Axis

(b)

Figure 8.24a and b   The Atlas and Axis, Cervical Vertebrae C1 and C2

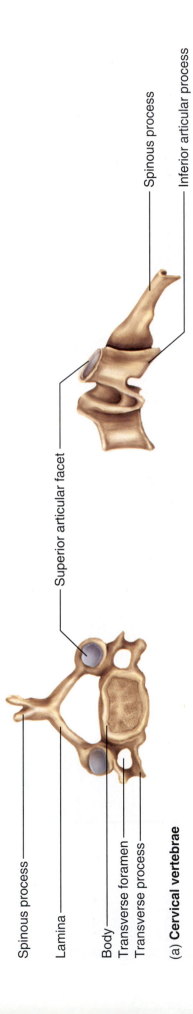

Spinous process

Inferior articular process

Superior articular facet

Spinous process

Lamina

Body

Transverse foramen

Transverse process

(a) **Cervical vertebrae**

Figure 8.25a Cervical Vertebra

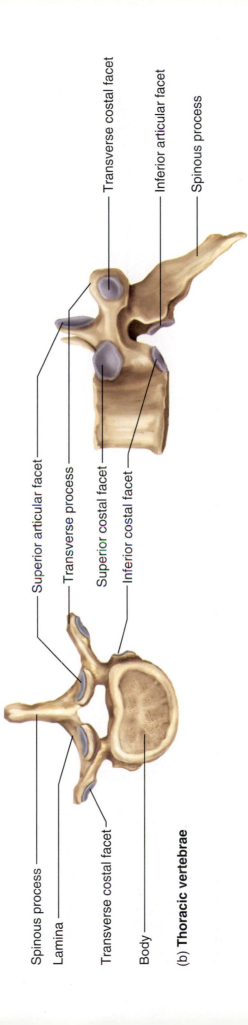

Transverse costal facet

Inferior articular facet

Spinous process

Superior articular facet

Transverse process

Superior costal facet

Inferior costal facet

Spinous process

Lamina

Transverse costal facet

Body

(b) **Thoracic vertebrae**

Figure 8.25b Thoracic Vertebrae

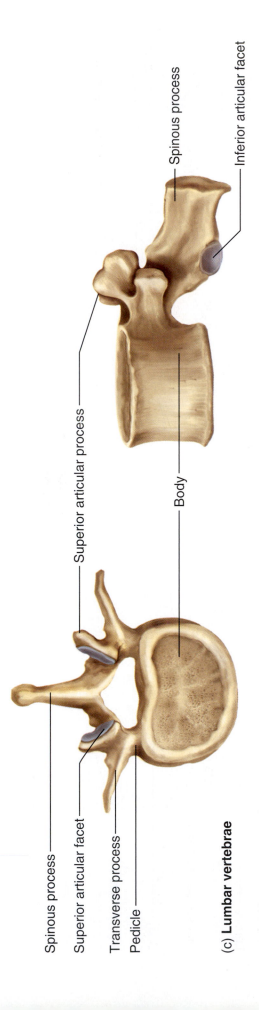

Spinous process

Inferior articular facet

Superior articular process

Body

Spinous process

Superior articular facet

Transverse process

Pedicle

(c) **Lumbar vertebrae**

Figure 8.25c   Lumbar Vertebrae

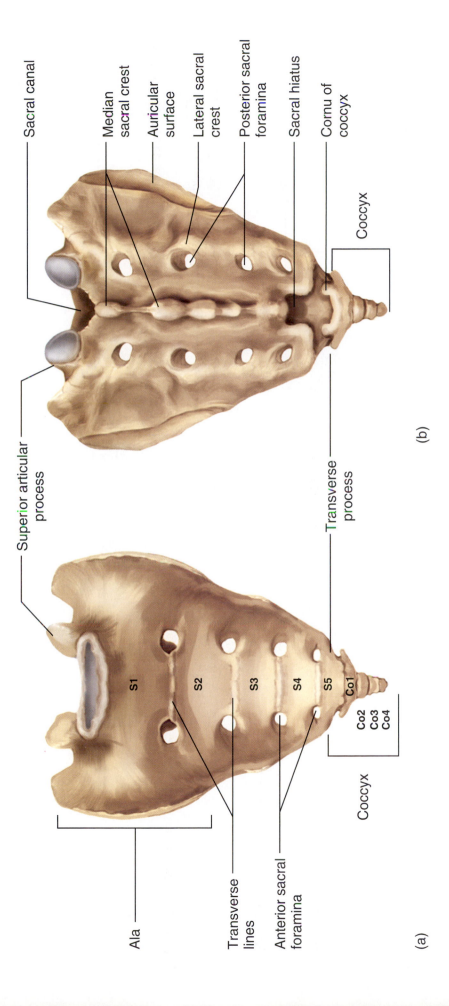

Sacral canal

Median sacral crest

Auricular surface

Lateral sacral crest

Posterior sacral foramina

Sacral hiatus

Cornu of coccyx

Coccyx

Superior articular process

Transverse process

Ala

Transverse lines

Anterior sacral foramina

Coccyx

S1

S2

S3

S4

S5

Co1

Co2
Co3
Co4

(a)

(b)

Figure 8.26 The Sacrum and Coccyx

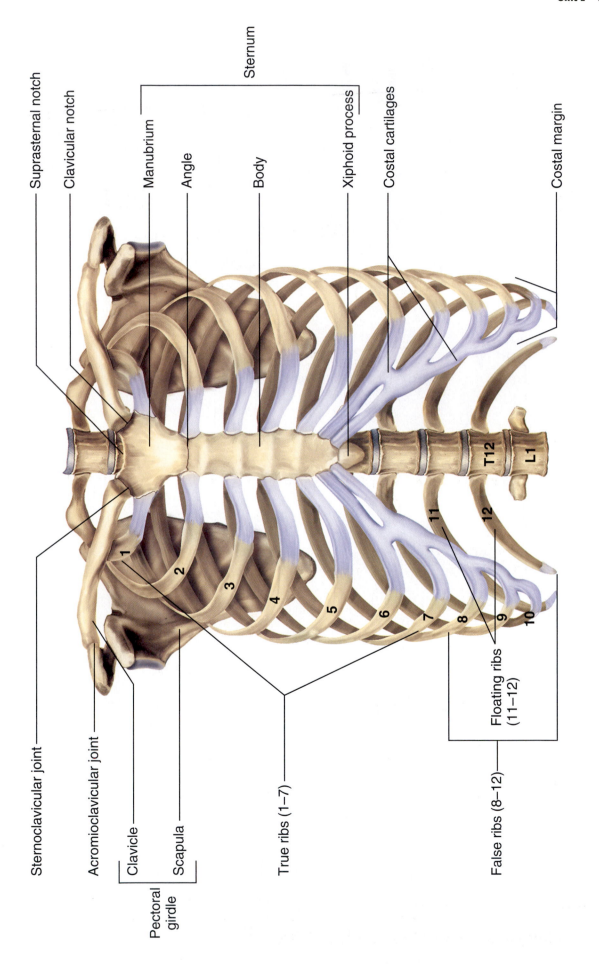

Suprasternal notch

Clavicular notch

Sternum

Manubrium

Angle

Body

Xiphoid process

Costal cartilages

Costal margin

Sternoclavicular joint

Acromioclavicular joint

Pectoral girdle { Clavicle, Scapula

True ribs (1–7)

False ribs (8–12)

Floating ribs (11–12)

T12

L1

1 2 3 4 5 6 7 8 9 10 11 12

Figure 8.27 The Thoracic Cage and Pectoral Girdle, Anterior View

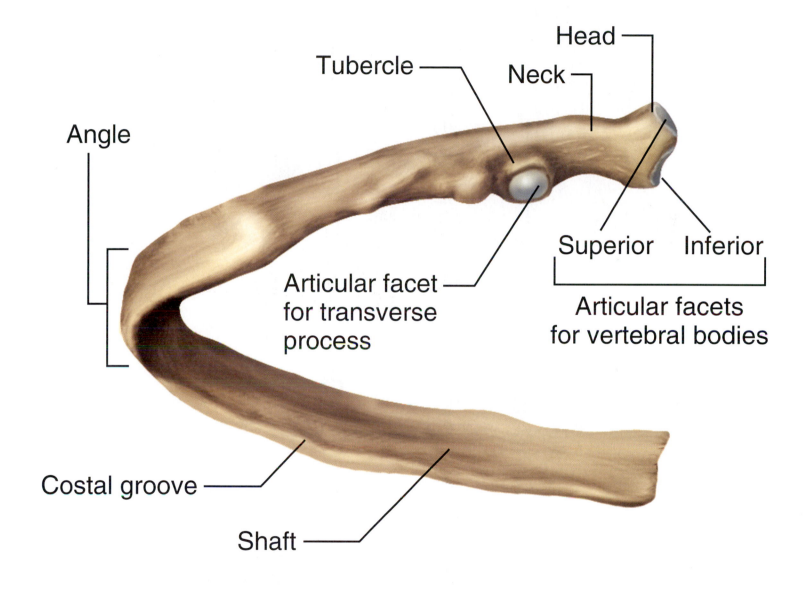

Head

Tubercle

Neck

Angle

Superior Inferior

Articular facet
for transverse
process

Articular facets
for vertebral bodies

Costal groove

Shaft

Figure 8.28b Anatomy of the Ribs—Typical Features of Ribs 2 to 10

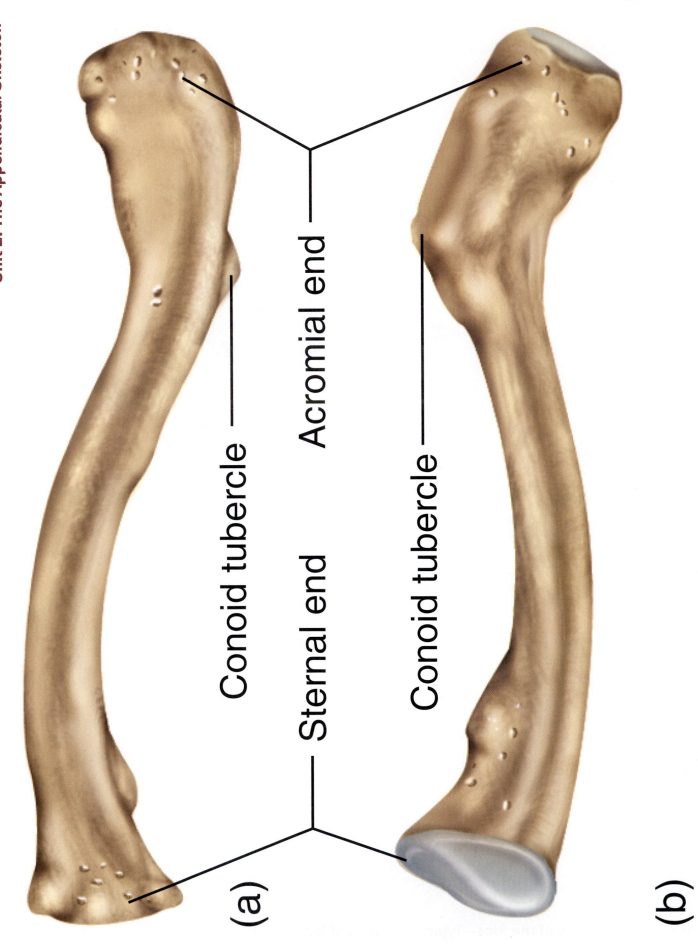

Conoid tubercle

Acromial end

Sternal end

Conoid tubercle

(a)

(b)

Figure 8.30 The Right Clavicle

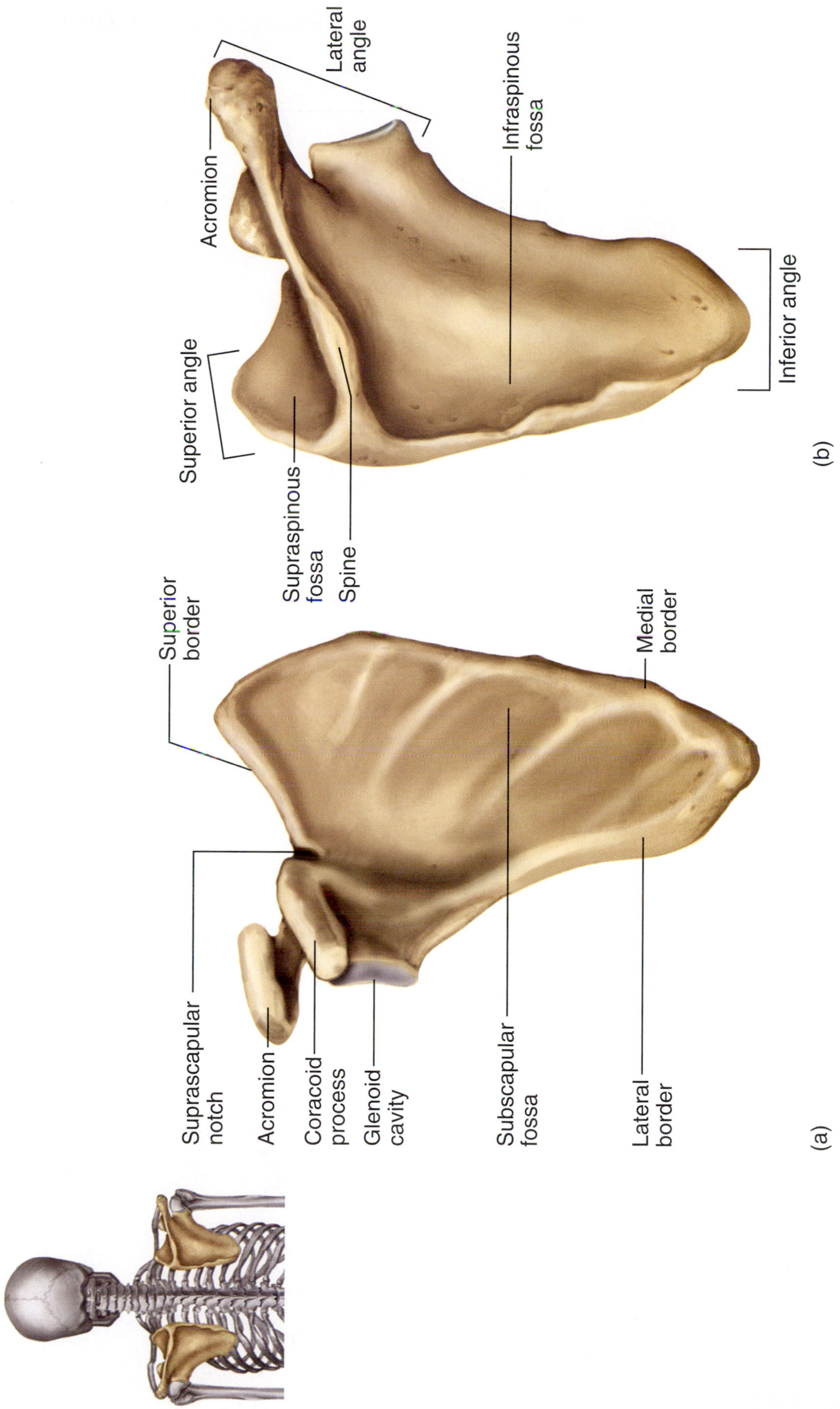

Lateral angle

Infraspinous fossa

Acromion

Inferior angle

Superior angle

Supraspinous fossa

Spine

(b)

Superior border

Medial border

Suprascapular notch

Acromion

Coracoid process

Glenoid cavity

Subscapular fossa

Lateral border

(a)

Figure 8.31 The Right Scapula

Anterior surface   **Posterior surface**

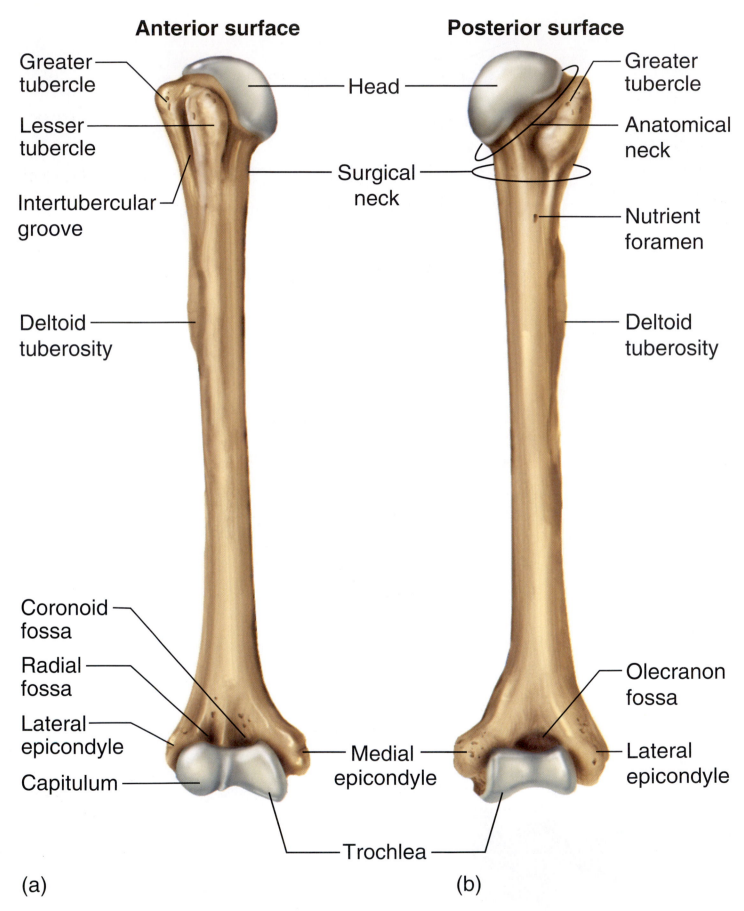

Greater
tubercle

Head

Greater
tubercle

Lesser
tubercle

Anatomical
neck

Intertubercular
groove

Surgical
neck

Nutrient
foramen

Deltoid
tuberosity

Deltoid
tuberosity

Coronoid
fossa

Radial
fossa

Olecranon
fossa

Lateral
epicondyle

Medial
epicondyle

Lateral
epicondyle

Capitulum

Trochlea

(a)   (b)

Figure 8.32   The Right Humerus

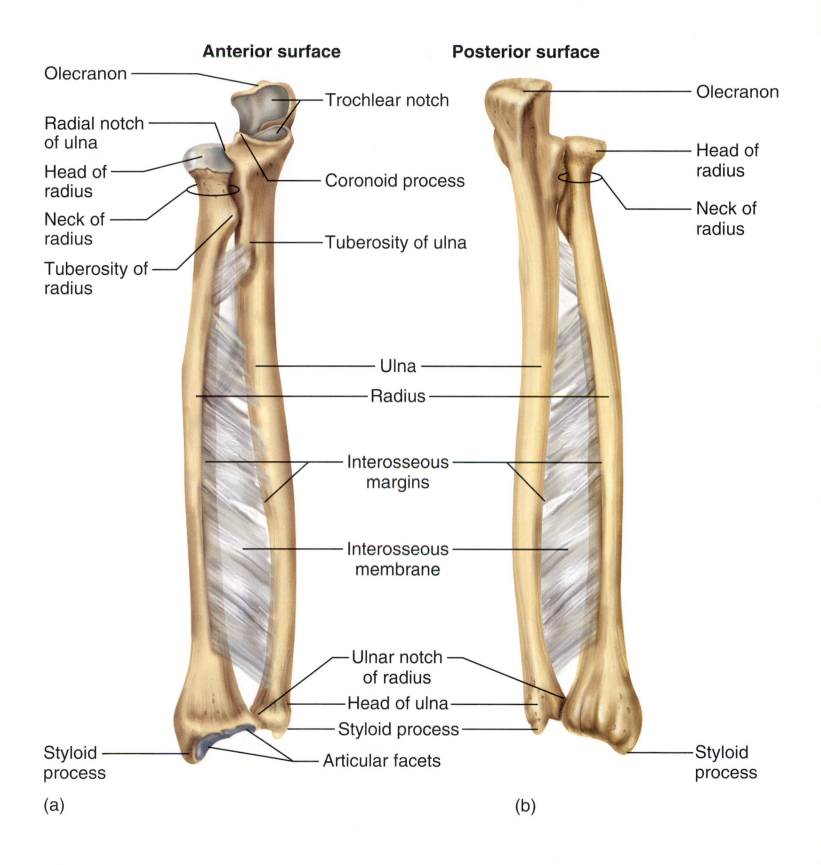

Anterior surface

Olecranon

Radial notch
of ulna

Head of
radius

Neck of
radius

Tuberosity of
radius

Trochlear notch

Coronoid process

Tuberosity of ulna

Posterior surface

Olecranon

Head of
radius

Neck of
radius

Ulna

Radius

Interosseous
margins

Interosseous
membrane

Ulnar notch
of radius

Head of ulna

Styloid process

Articular facets

Styloid
process

Styloid
process

(a)

(b)

Figure 8.33 **The Right Radius and Ulna**

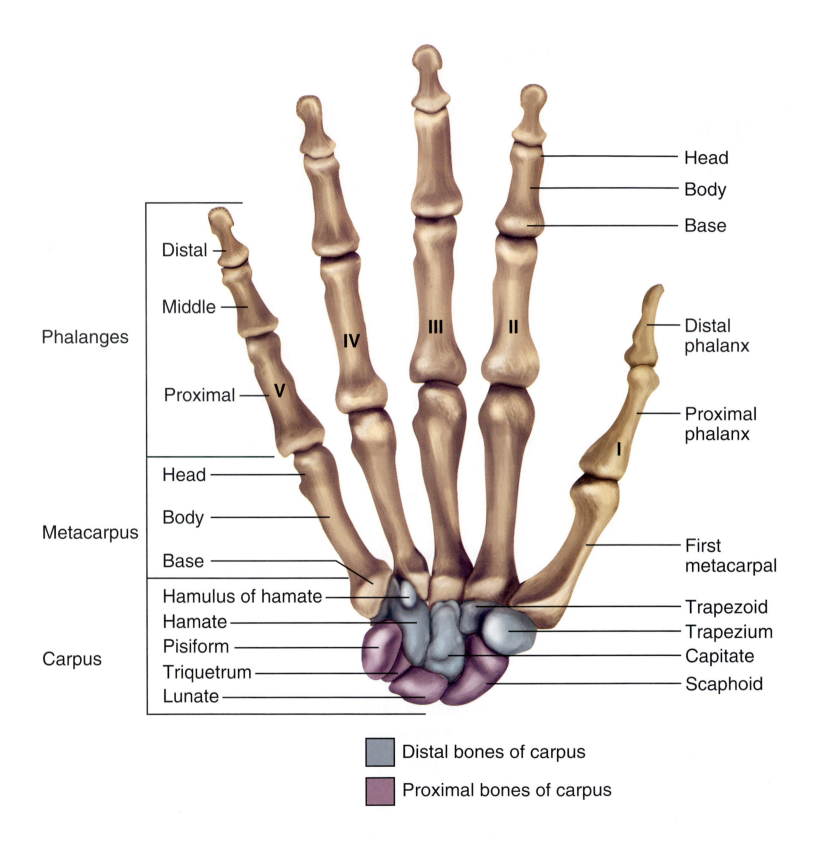

Head

Body

Base

Phalanges

Distal

Middle

Distal
phalanx

IV

III

II

Proximal

V

Proximal
phalanx

I

Head

Metacarpus

Body

Base

First
metacarpal

Hamulus of hamate

Hamate

Trapezoid

Pisiform

Trapezium

Carpus

Triquetrum

Capitate

Lunate

Scaphoid

Distal bones of carpus

Proximal bones of carpus

Figure 8.34a   **The Right Wrist and Hand, Anterior View**

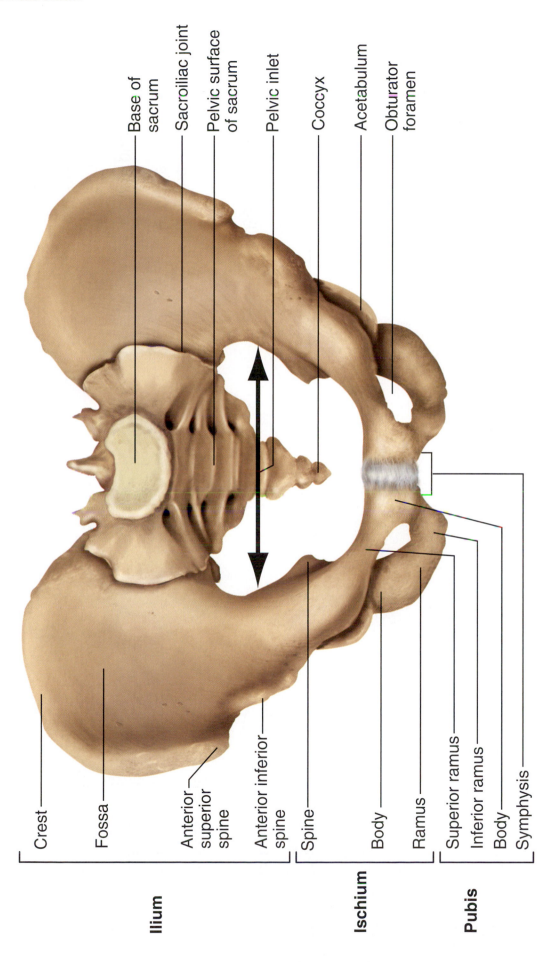

Base of sacrum

Sacroiliac joint

Pelvic surface of sacrum

Pelvic inlet

Coccyx

Acetabulum

Obturator foramen

Crest

Fossa

Anterior superior spine

Anterior inferior spine

Spine

Body

Ramus

Superior ramus

Inferior ramus

Body

Symphysis

Ilium

Ischium

Pubis

Figure 8.35   The Pelvic Girdle, Anterosuperior View

Iliac crest

Anterior gluteal line

Anterior superior spine of ilium

Anterior inferior spine of ilium

Body of ilium

Superior ramus of pubis

Body of pubis

Inferior ramus of pubis

Obturator foramen

Ramus of ischium

Pubis

Ischium

Ilium

Inferior gluteal line

Posterior gluteal line

Posterior superior spine of ilium

Posterior inferior spine of ilium

Greater sciatic notch

Acetabulum

Spine of ischium

Lesser sciatic notch

Body of ischium

Ischial tuberosity

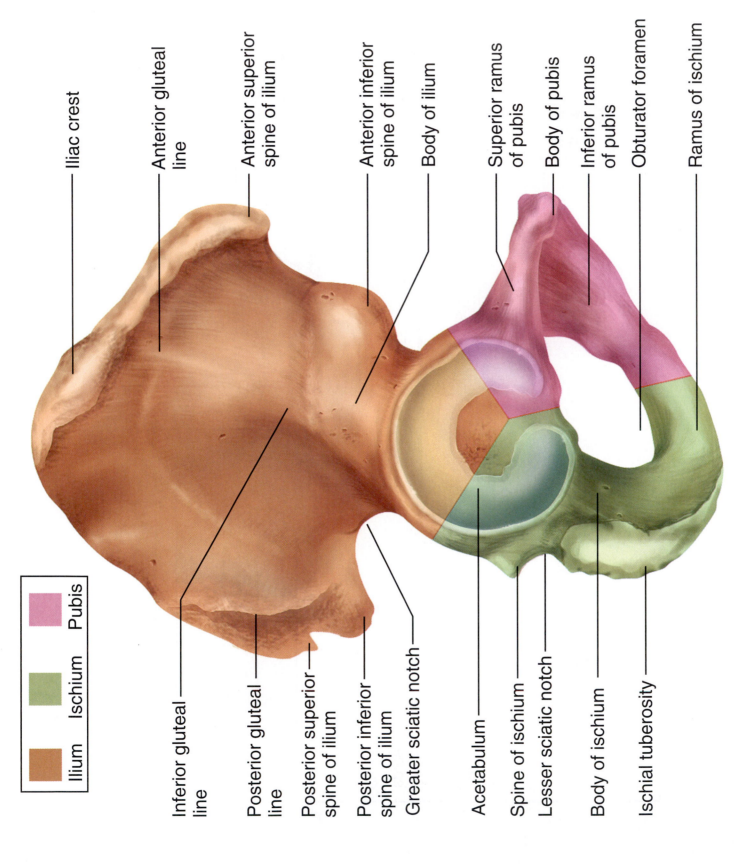

Figure 8.36   The Right Os Coxae, Lateral View

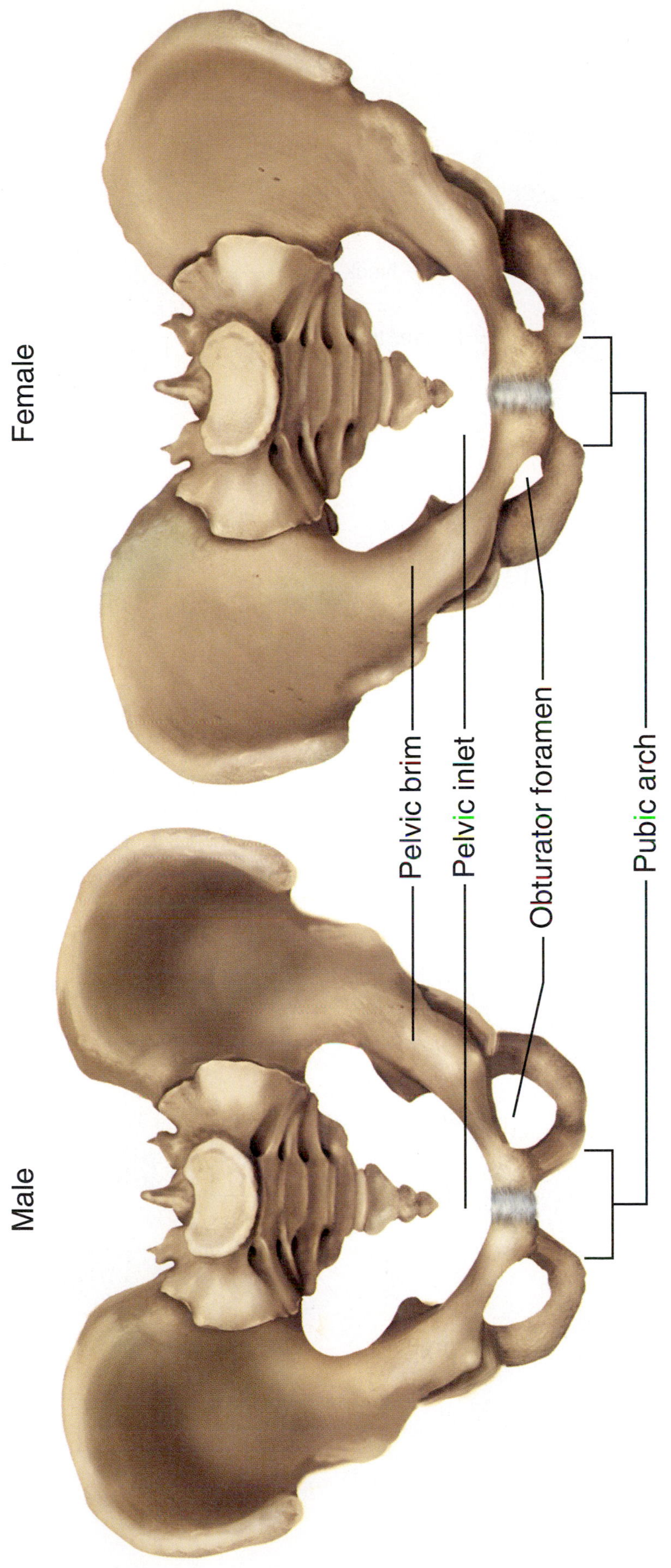

Female

Male

Pelvic brim

Pelvic inlet

Obturator foramen

Pubic arch

Figure 8.37 Comparison of the Male and Female Pelvic Girdle

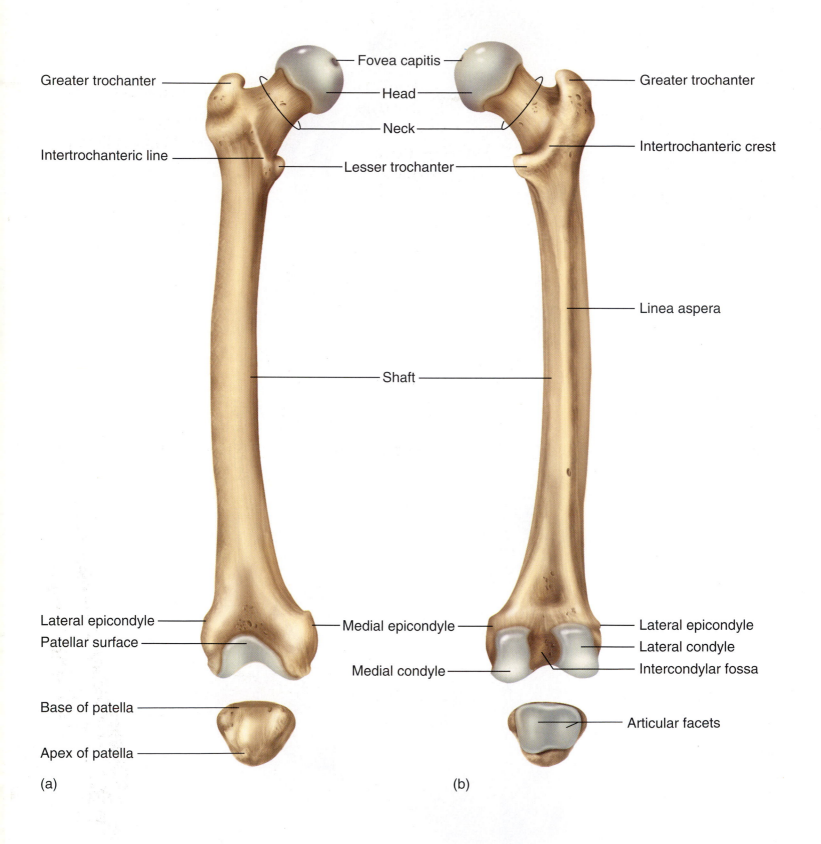

Greater trochanter

Intertrochanteric line

Lateral epicondyle

Patellar surface

Base of patella

Apex of patella

(a)

Fovea capitis

Head

Neck

Lesser trochanter

Shaft

Medial epicondyle

Medial condyle

Greater trochanter

Intertrochanteric crest

Linea aspera

Lateral epicondyle

Lateral condyle

Intercondylar fossa

Articular facets

(b)

Figure 8.38   **The Right Femur and Patella**

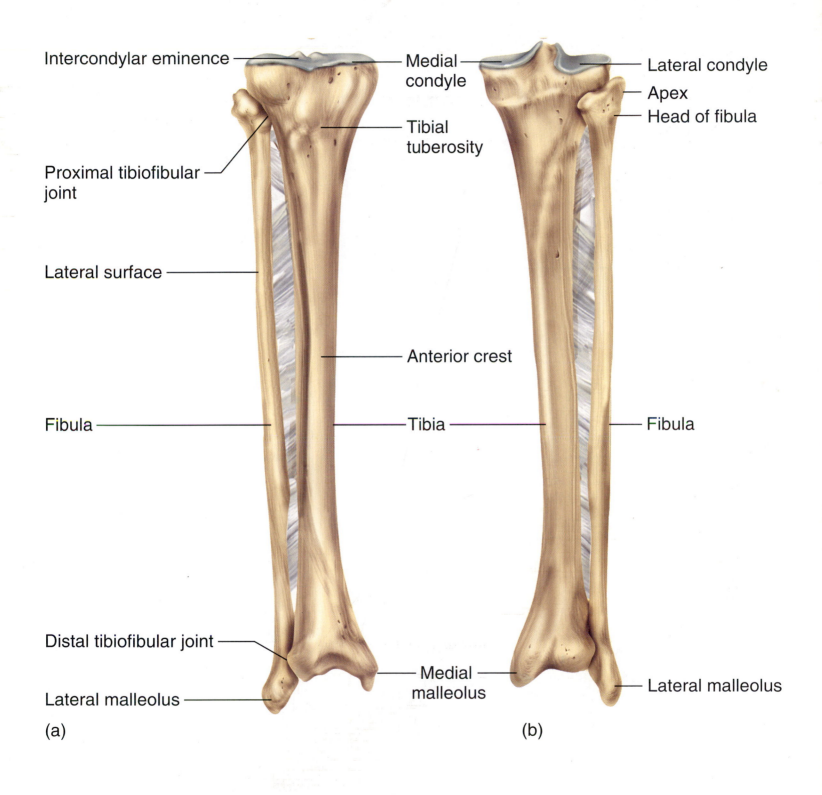

Intercondylar eminence

Medial condyle

Lateral condyle

Apex

Head of fibula

Tibial tuberosity

Proximal tibiofibular joint

Lateral surface

Anterior crest

Fibula

Tibia

Fibula

Distal tibiofibular joint

Medial malleolus

Lateral malleolus

Lateral malleolus

(a)

(b)

Figure 8.39 **The Right Tibia and Fibula**

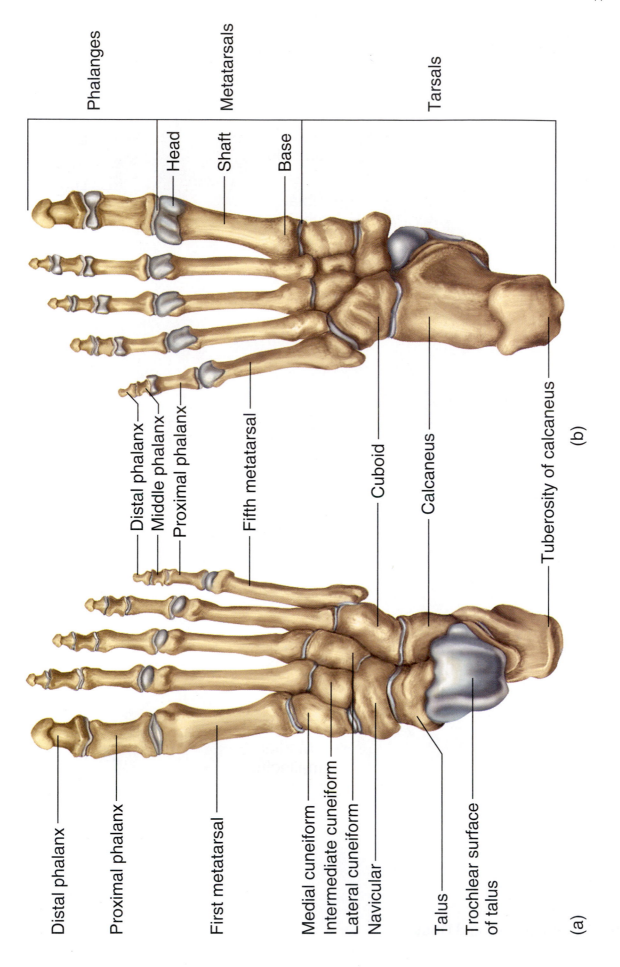

Phalanges
Metatarsals
Head
Shaft
Base
Tarsals

Distal phalanx
Middle phalanx
Proximal phalanx
Fifth metatarsal
Cuboid
Calcaneus
Tuberosity of calcaneus
(b)

Distal phalanx
Proximal phalanx
First metatarsal
Medial cuneiform
Intermediate cuneiform
Lateral cuneiform
Navicular
Talus
Trochlear surface of talus
(a)

Figure 8.40 The Right Foot

Unit 3: Articulations

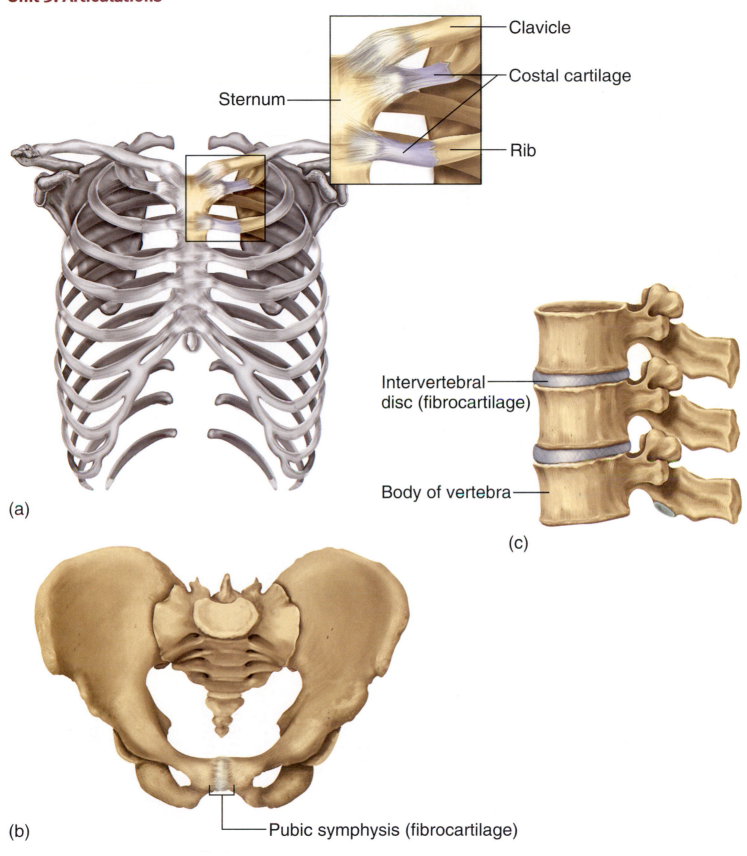

(a)

Clavicle

Costal cartilage

Sternum

Rib

Intervertebral disc (fibrocartilage)

Body of vertebra

(c)

(b)

Pubic symphysis (fibrocartilage)

Figure 9.5   Cartilaginous Joints

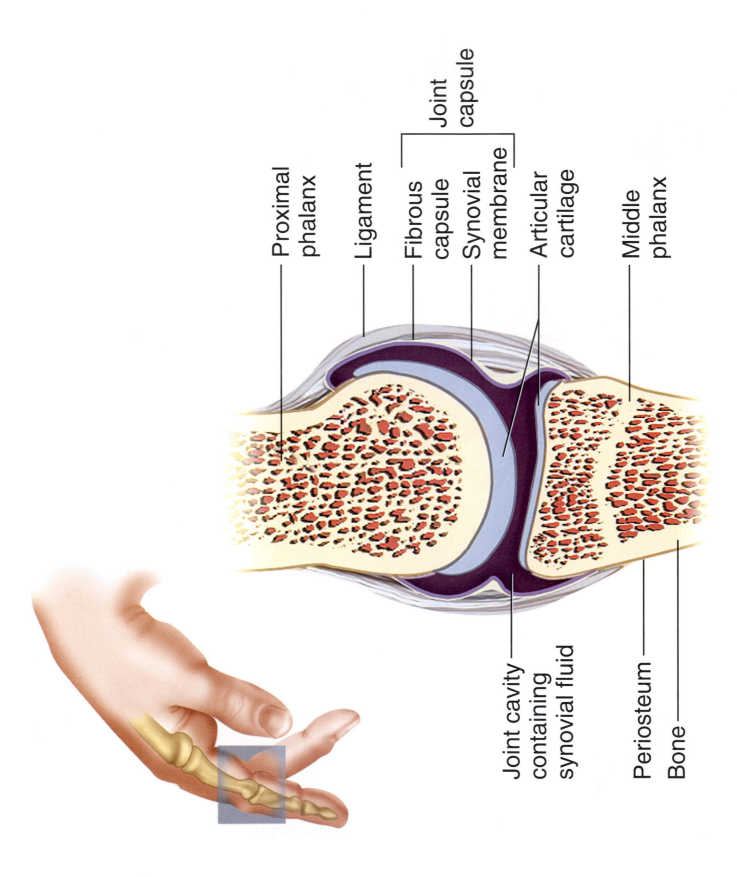

Proximal phalanx

Ligament

Joint capsule
{ Fibrous capsule
  Synovial membrane }

Articular cartilage

Middle phalanx

Joint cavity containing synovial fluid

Periosteum

Bone

Figure 9.6   Structure of a Simple Synovial Joint

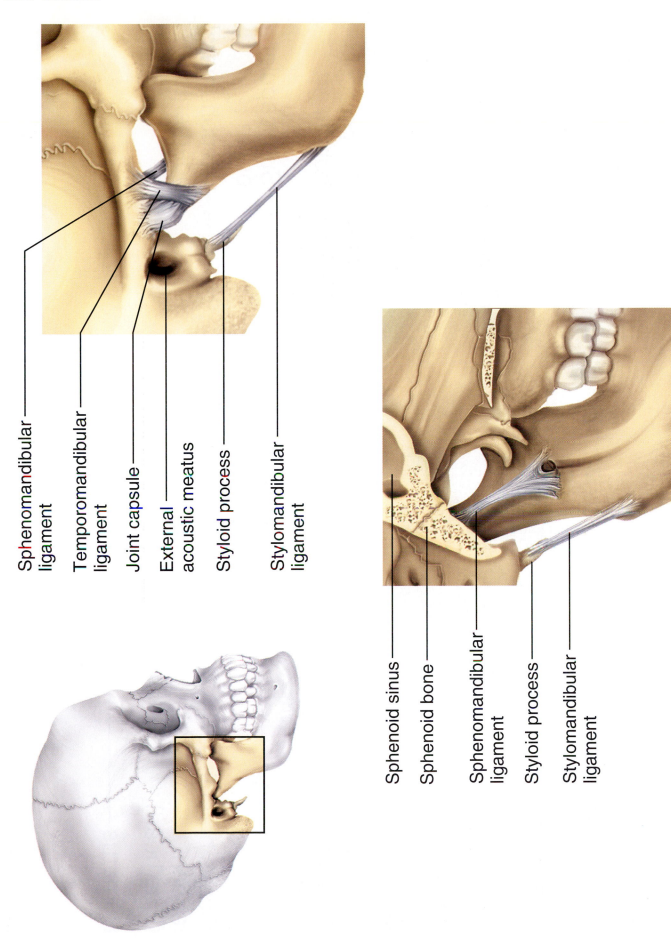

Sphenomandibular ligament

Temporomandibular ligament

Joint capsule

External acoustic meatus

Styloid process

Stylomandibular ligament

Sphenoid sinus

Sphenoid bone

Sphenomandibular ligament

Styloid process

Stylomandibular ligament

Figure 9.18a and b The Temporomandibular Joint (TMJ)

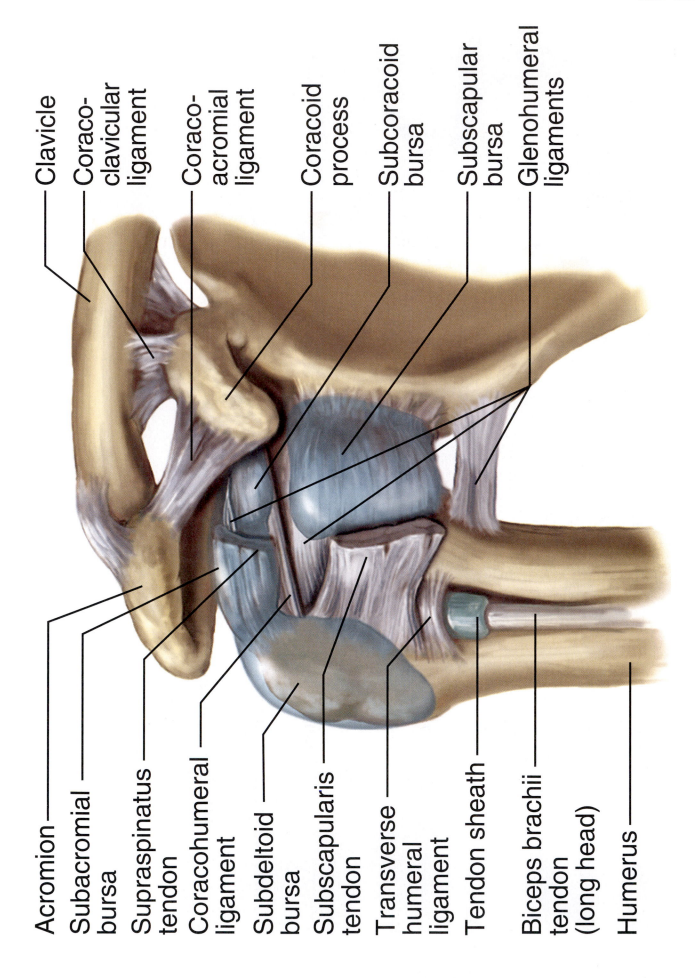

Clavicle

Coraco-
clavicular
ligament

Coraco-
acromial
ligament

Coracoid
process

Subcoracoid
bursa

Subscapular
bursa

Glenohumeral
ligaments

Acromion

Subacromial
bursa

Supraspinatus
tendon

Coracohumeral
ligament

Subdeltoid
bursa

Subscapularis
tendon

Transverse
humeral
ligament

Tendon sheath

Biceps brachii
tendon
(long head)

Humerus

Figure 9.19a The Humeroscapular Joint—Anterior View

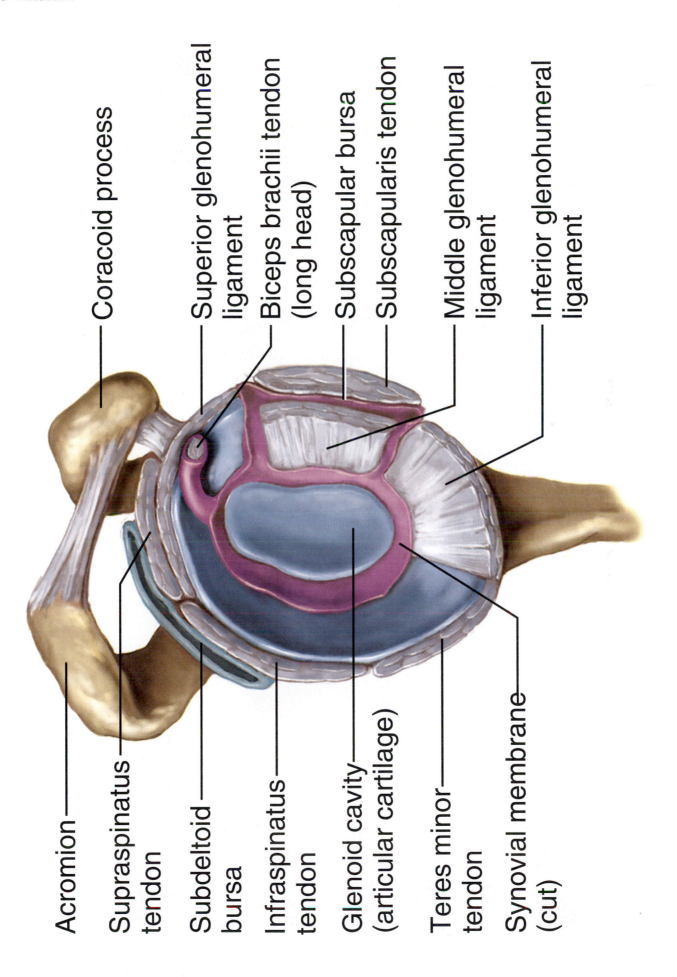

Acromion

Supraspinatus tendon

Subdeltoid bursa

Infraspinatus tendon

Glenoid cavity (articular cartilage)

Teres minor tendon

Synovial membrane (cut)

Coracoid process

Superior glenohumeral ligament

Biceps brachii tendon (long head)

Subscapular bursa

Subscapularis tendon

Middle glenohumeral ligament

Inferior glenohumeral ligament

Figure 9.19b The Humeroscapular Joint—Lateral View

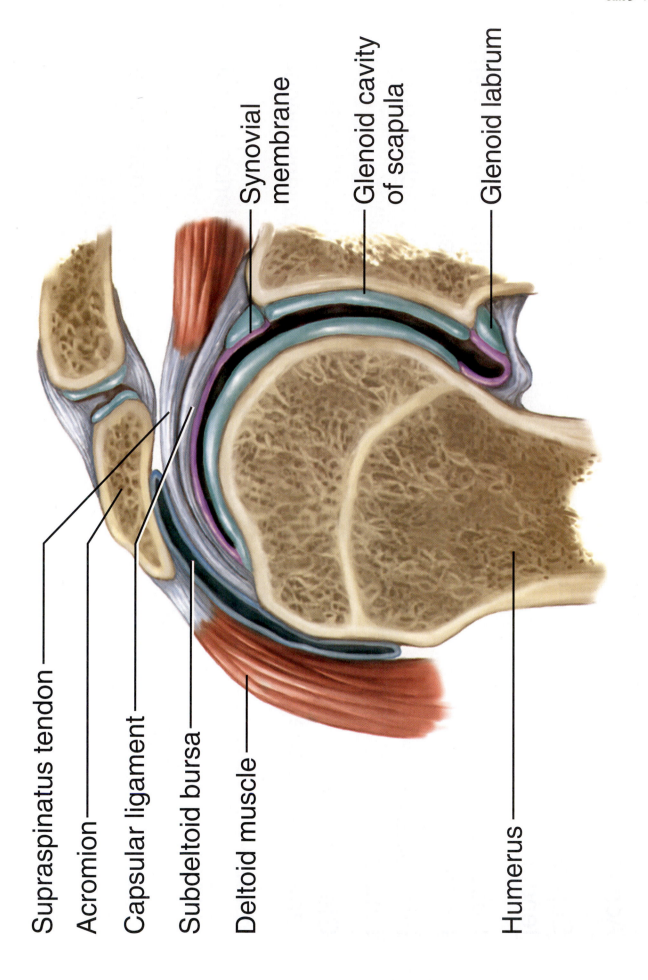

Supraspinatus tendon

Acromion

Capsular ligament

Subdeltoid bursa

Deltoid muscle

Synovial membrane

Glenoid cavity of scapula

Glenoid labrum

Humerus

Figure 9.19c Frontal Section of the Right Shoulder Joint, Anterior View

Acromion of scapula

Clavicle

Head of humerus

Coracobrachialis
muscle

Deltoid muscle
(cut and folded back)

Pectoralis major
muscle

Biceps brachii muscle
    Short head
    Long head

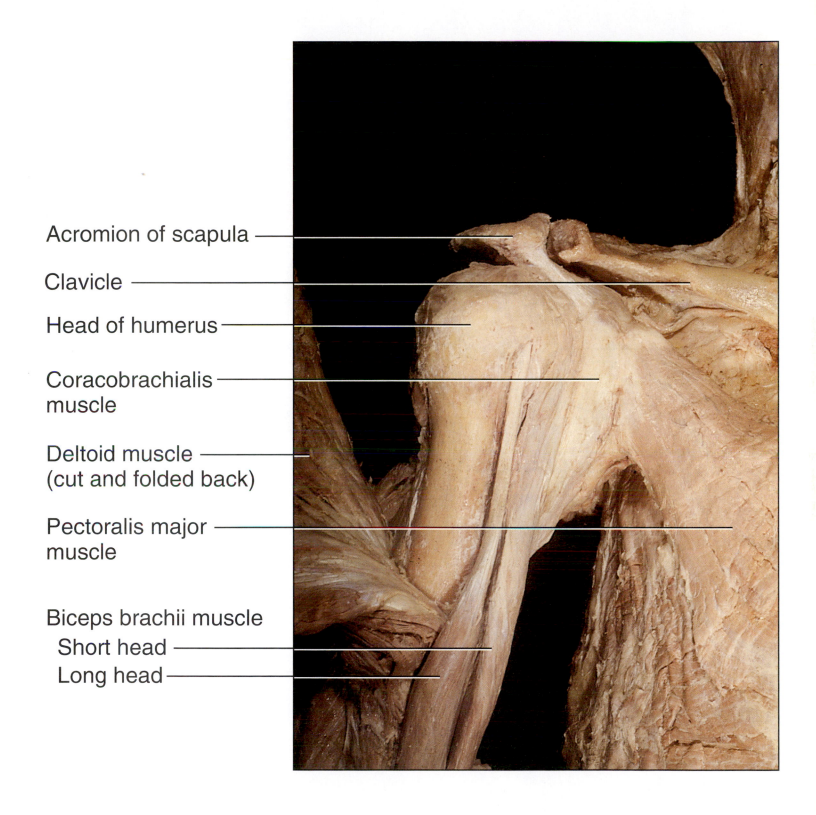

Figure 9.19d   Photograph of the Humeroscapular Joint

Anterior

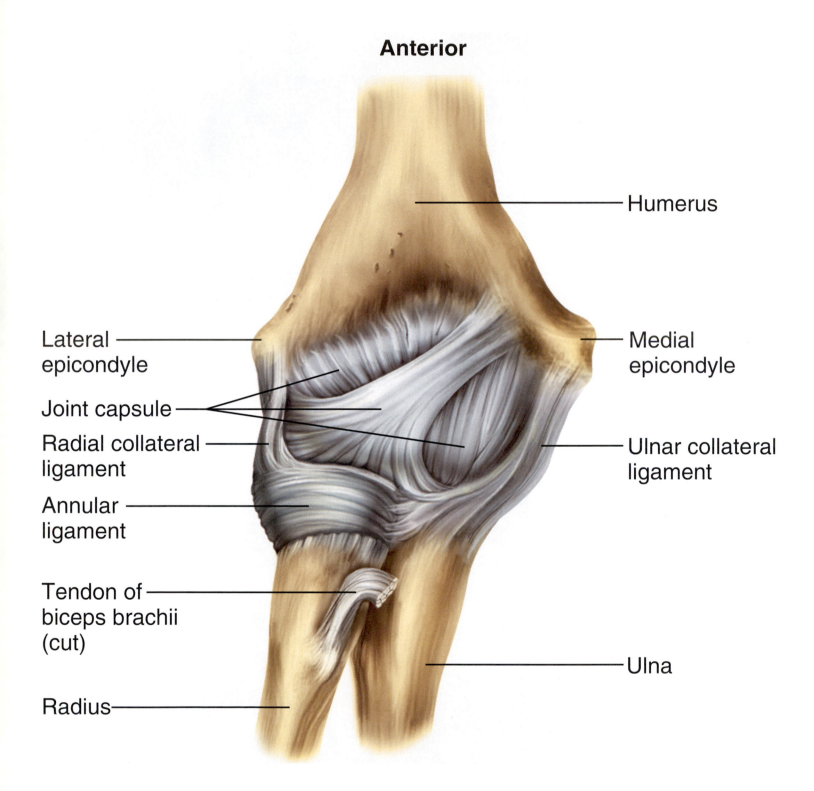

Humerus

Lateral
epicondyle

Medial
epicondyle

Joint capsule

Radial collateral
ligament

Ulnar collateral
ligament

Annular
ligament

Tendon of
biceps brachii
(cut)

Ulna

Radius

Figure 9.20a   The Elbow Joint—Anterior View

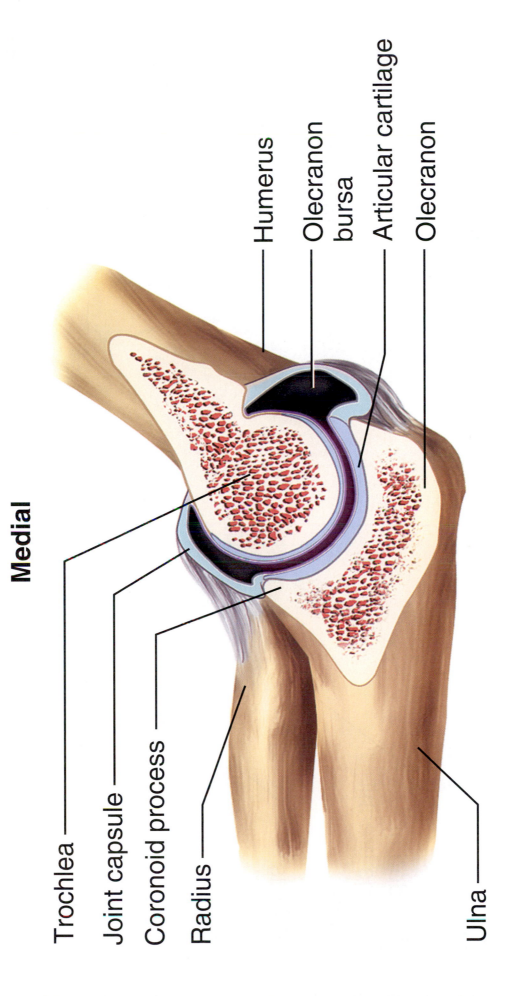

Medial

Humerus

Olecranon bursa

Articular cartilage

Olecranon

Trochlea

Joint capsule

Coronoid process

Radius

Ulna

Figure 9.20b The Elbow Joint—Midsaggital Section, Medial View

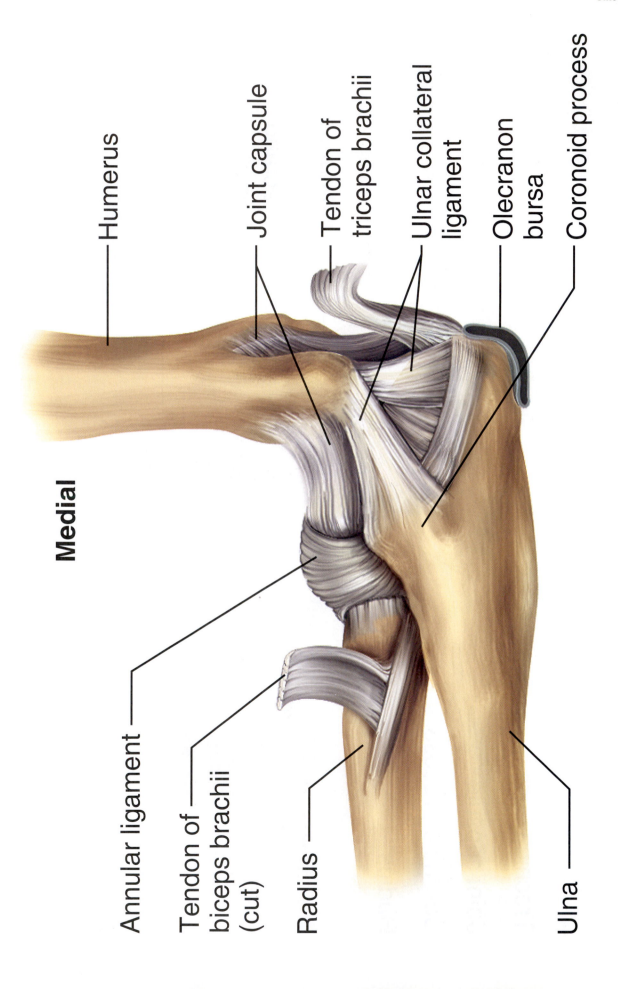

Humerus

Joint capsule

Tendon of
triceps brachii

Ulnar collateral
ligament

Olecranon
bursa

Coronoid process

Medial

Annular ligament

Tendon of
biceps brachii
(cut)

Radius

Ulna

Figure 9.20c The Elbow Joint—Medial View

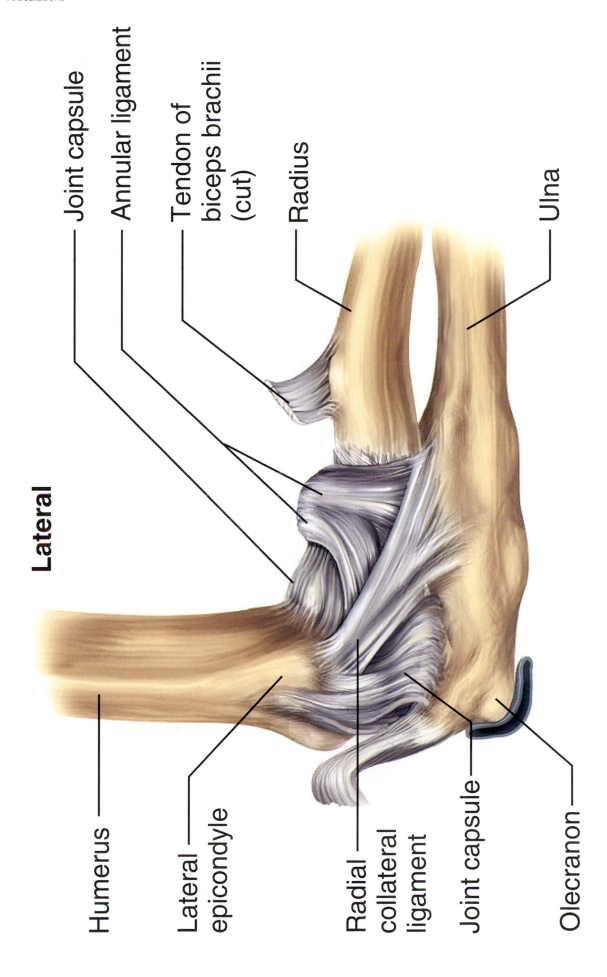

Joint capsule

Annular ligament

Tendon of biceps brachii (cut)

Radius

Ulna

Lateral

Humerus

Lateral epicondyle

Radial collateral ligament

Joint capsule

Olecranon

Figure 9.20d The Elbow Joint—Lateral View

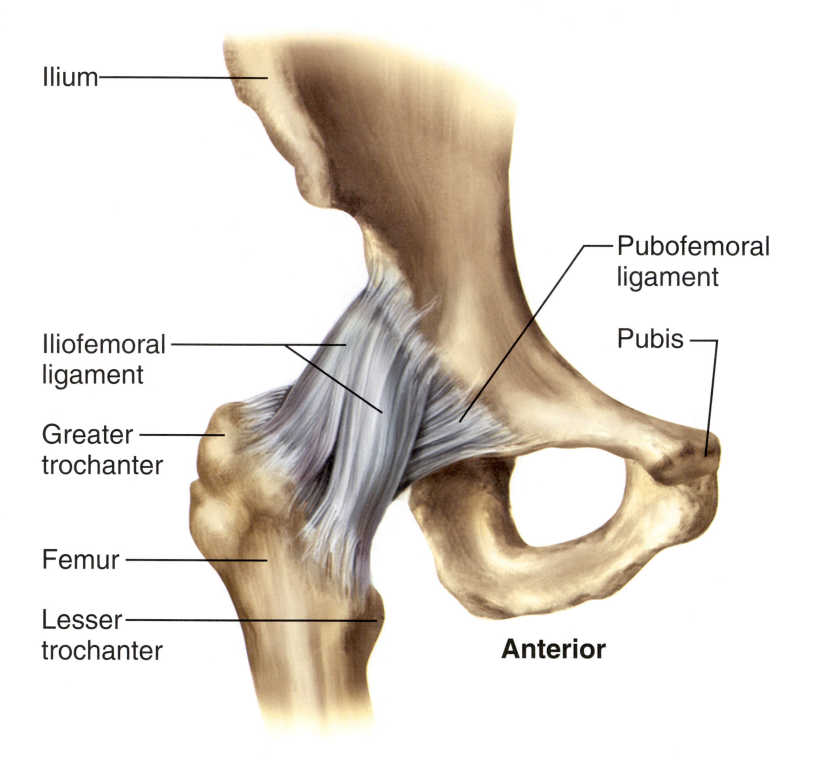

Ilium

Pubofemoral
ligament

Iliofemoral
ligament

Pubis

Greater
trochanter

Femur

Lesser
trochanter

Anterior

Figure 9.21a The Coxal Joint—Anterior View

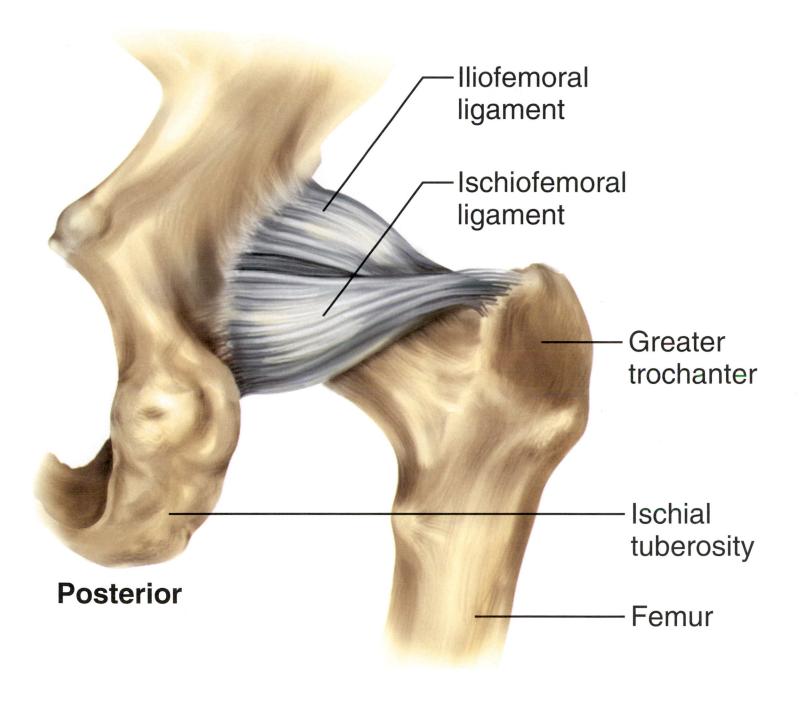

Iliofemoral ligament

Ischiofemoral ligament

Greater trochanter

Ischial tuberosity

Femur

Posterior

Figure 9.21b The Coxal Joint—Posterior View

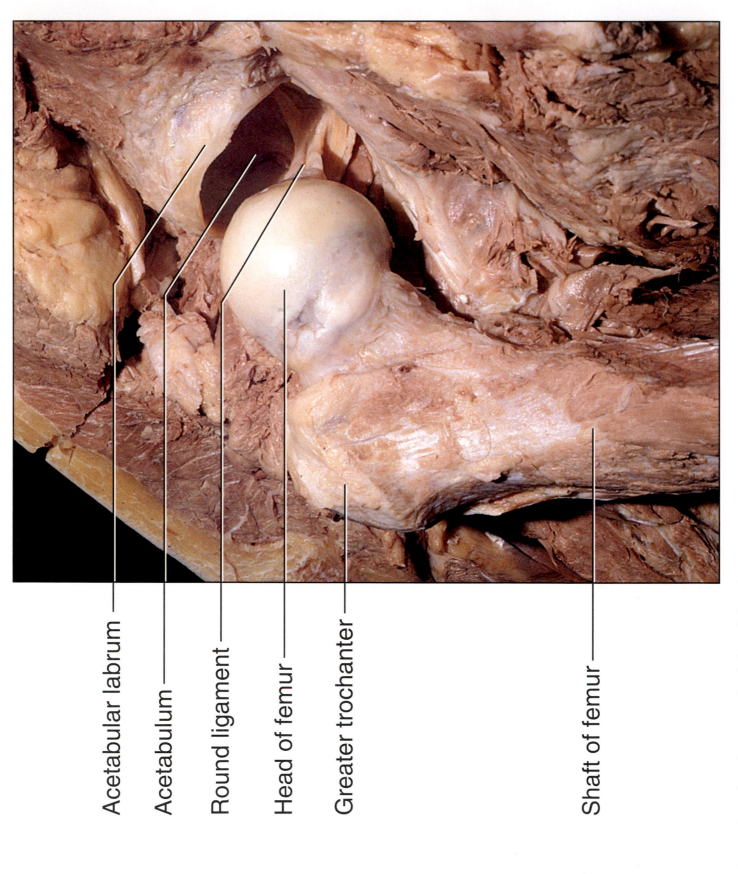

Acetabular labrum

Acetabulum

Round ligament

Head of femur

Greater trochanter

Shaft of femur

Figure 9.21d   Photograph of the Right Hip

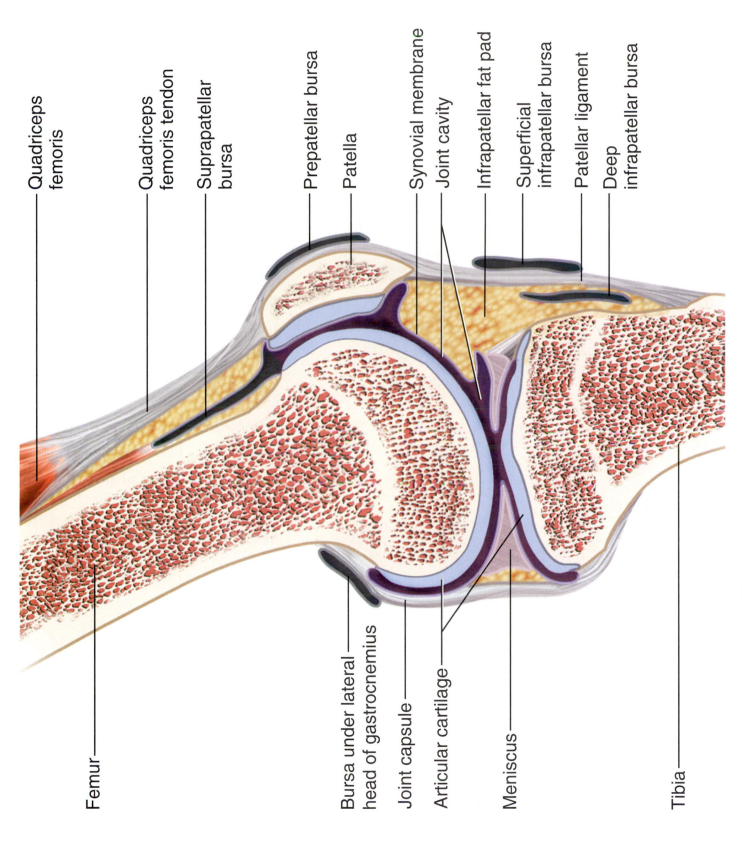

Quadriceps femoris

Quadriceps femoris tendon

Suprapatellar bursa

Prepatellar bursa

Patella

Synovial membrane

Joint cavity

Infrapatellar fat pad

Superficial infrapatellar bursa

Patellar ligament

Deep infrapatellar bursa

Femur

Bursa under lateral head of gastrocnemius

Joint capsule

Articular cartilage

Meniscus

Tibia

Figure 9.23a The Tibiofemoral Joint—Midsaggital Section

Anterior

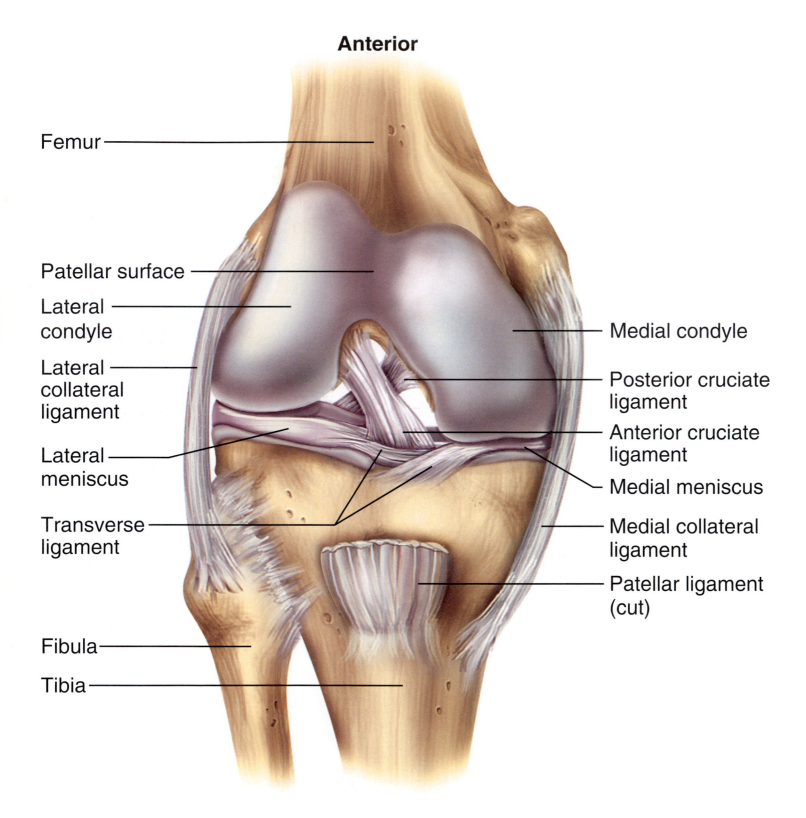

Femur

Patellar surface

Lateral condyle

Lateral collateral ligament

Lateral meniscus

Transverse ligament

Fibula

Tibia

Medial condyle

Posterior cruciate ligament

Anterior cruciate ligament

Medial meniscus

Medial collateral ligament

Patellar ligament (cut)

Figure 9.23b The Tibiofemoral Joint—Anterior View

Posterior

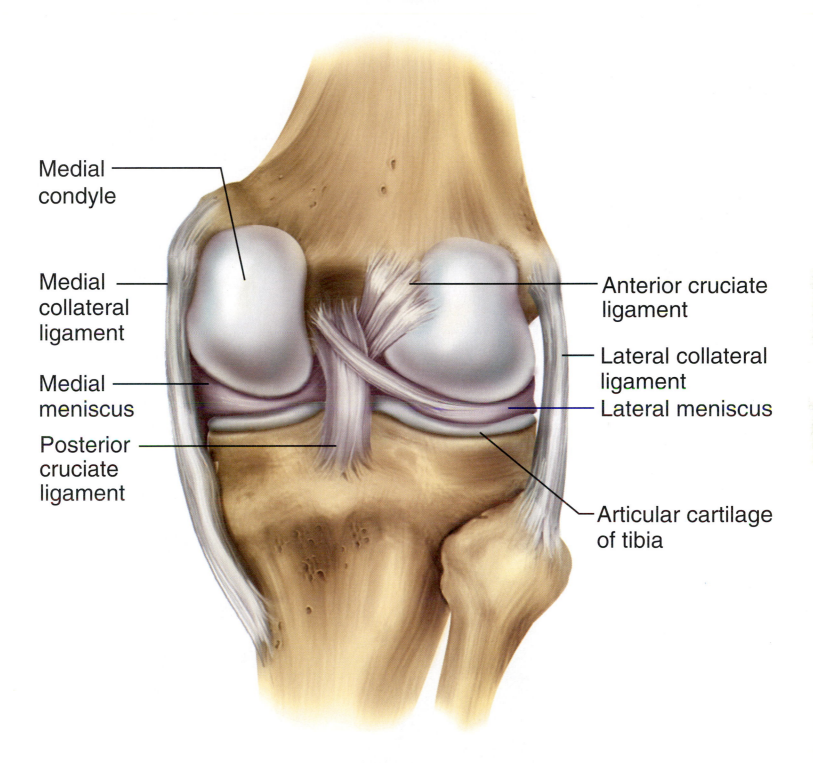

Medial
condyle

Medial
collateral
ligament

Medial
meniscus

Posterior
cruciate
ligament

Anterior cruciate
ligament

Lateral collateral
ligament

Lateral meniscus

Articular cartilage
of tibia

Figure 9.23c   The Tibiofemoral Joint—Posterior View

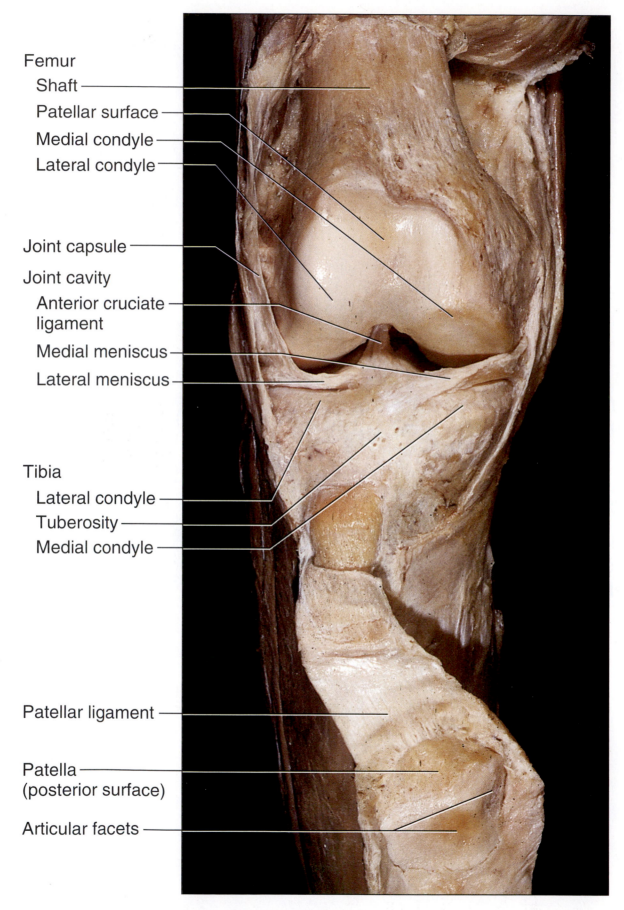

Femur
  Shaft
  Patellar surface
  Medial condyle
  Lateral condyle

Joint capsule

Joint cavity
  Anterior cruciate
  ligament
  Medial meniscus
  Lateral meniscus

Tibia
  Lateral condyle
  Tuberosity
  Medial condyle

Patellar ligament

Patella
(posterior surface)

Articular facets

Figure 9.24   Photograph of the Knee Joint

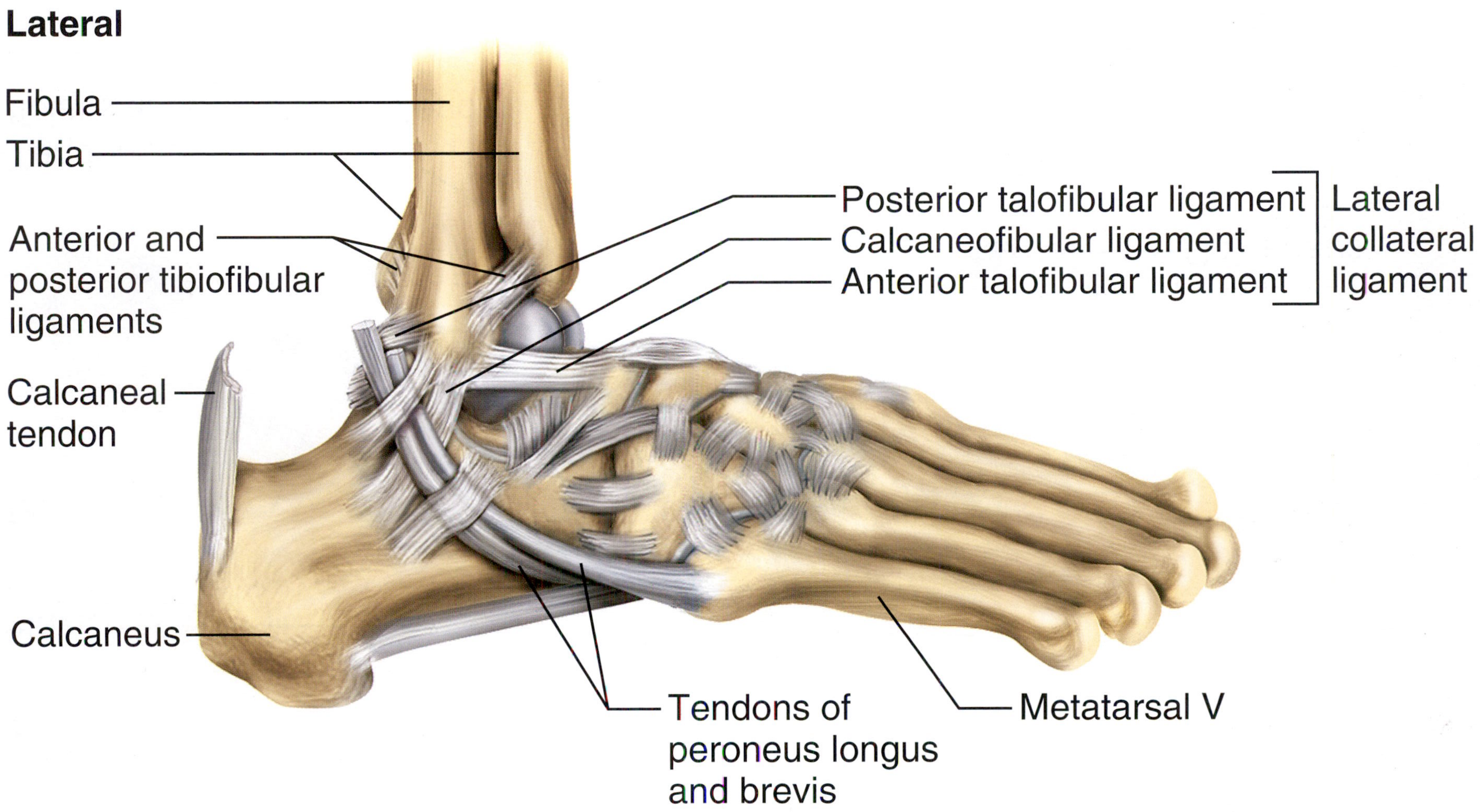

Lateral

Fibula

Tibia

Anterior and
posterior tibiofibular
ligaments

Calcaneal
tendon

Calcaneus

Posterior talofibular ligament
Calcaneofibular ligament
Anterior talofibular ligament

Lateral
collateral
ligament

Metatarsal V

Tendons of
peroneus longus
and brevis

Figure 9.25a The Talocrural Joint and Ligaments of the Right Foot—Lateral View

Medial

Deltoid ligament

Navicular

Metatarsal I

Tibia

Calcaneal tendon

Calcaneus

Tendons of tibialis anterior and posterior

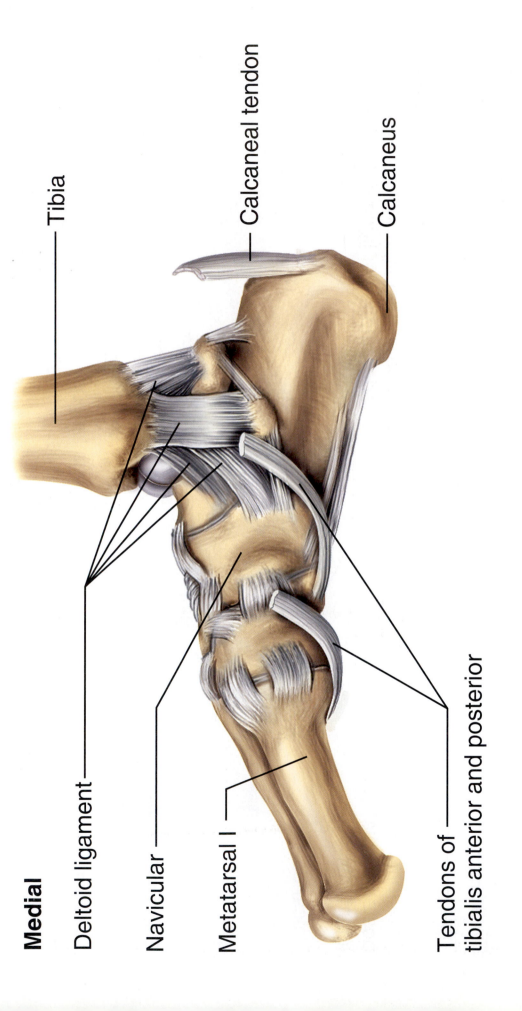

Figure 9.25b The Talocrural Joint and Ligaments of the Right Foot—Medial View

Posterior

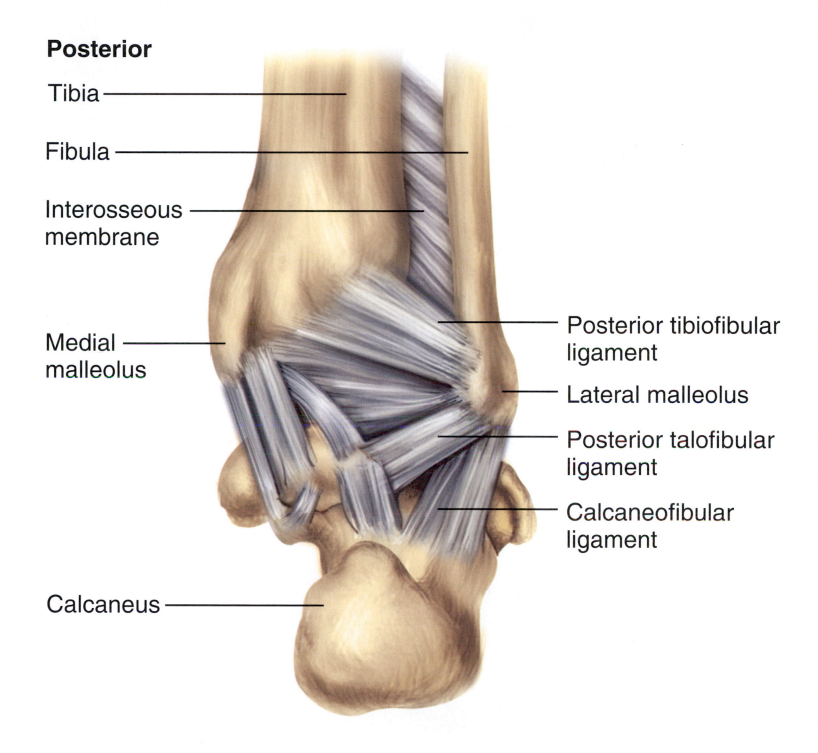

Tibia

Fibula

Interosseous membrane

Medial malleolus

Posterior tibiofibular ligament

Lateral malleolus

Posterior talofibular ligament

Calcaneofibular ligament

Calcaneus

Figure 9.25c The Talocrural Joint and Ligaments of the Right Foot—Posterior View

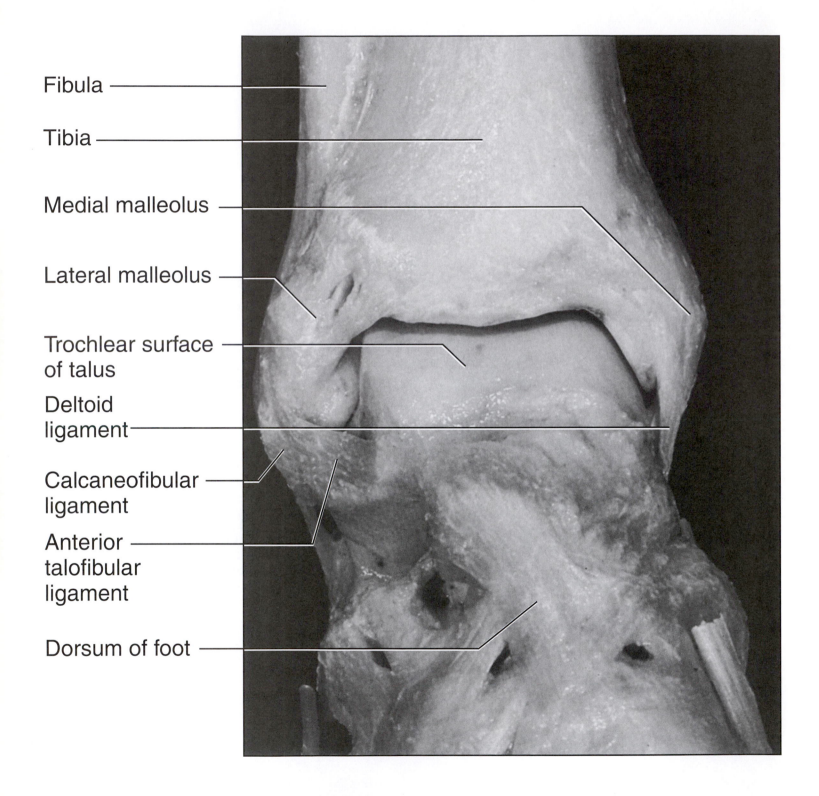

Fibula

Tibia

Medial malleolus

Lateral malleolus

Trochlear surface
of talus

Deltoid
ligament

Calcaneofibular
ligament

Anterior
talofibular
ligament

Dorsum of foot

Figure 9.26   Photograph of the Talocrural Joint, Anterior Joint

Unit 4: The Muscular System

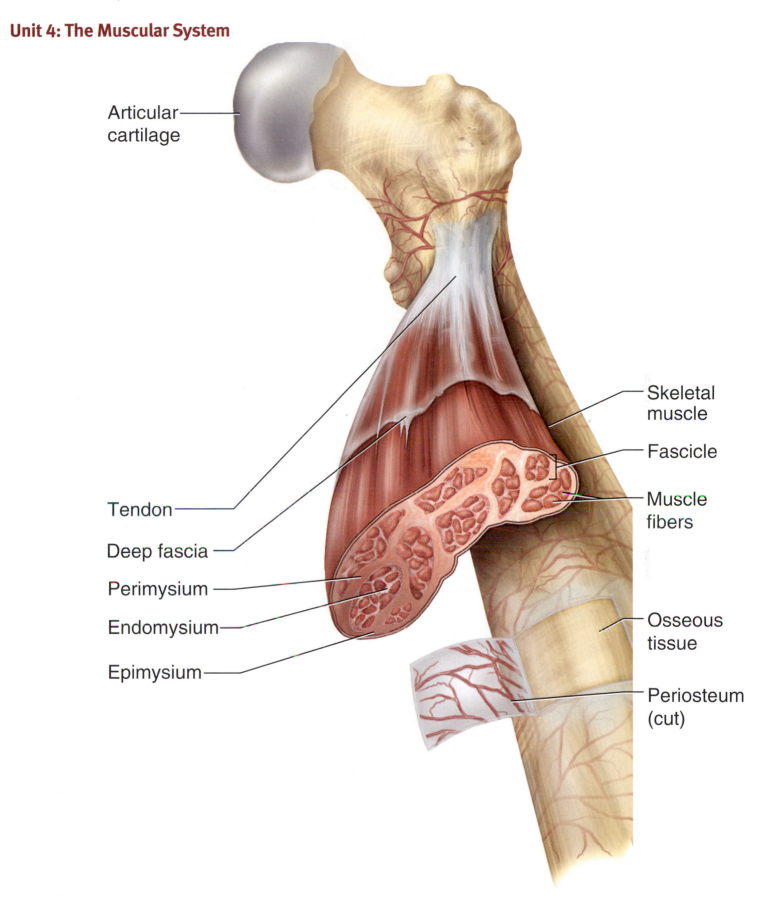

Articular cartilage

Skeletal muscle

Fascicle

Tendon

Muscle fibers

Deep fascia

Perimysium

Endomysium

Osseous tissue

Epimysium

Periosteum (cut)

Figure 10.1a   **Connective Tissues of a Muscle**

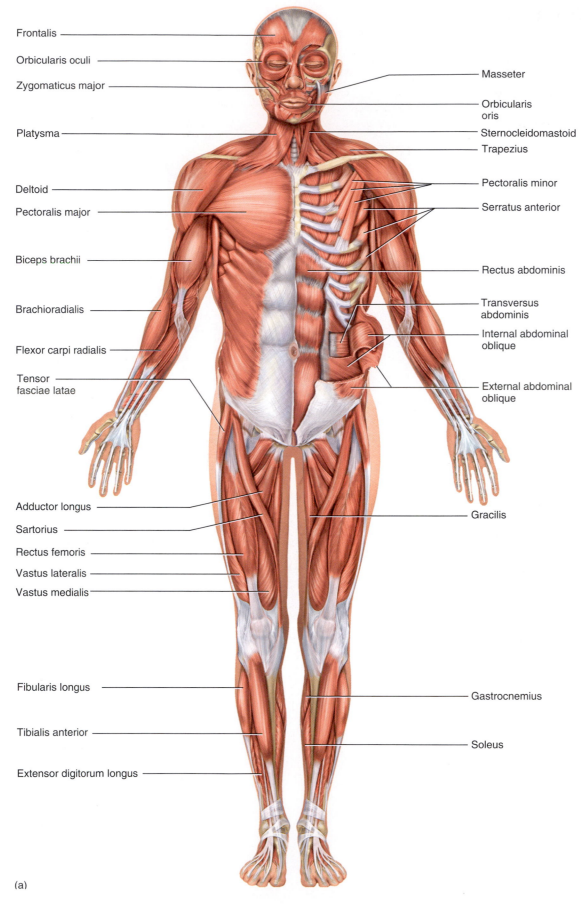

Frontalis

Orbicularis oculi

Zygomaticus major

Platysma

Deltoid

Pectoralis major

Biceps brachii

Brachioradialis

Flexor carpi radialis

Tensor
fasciae latae

Adductor longus

Sartorius

Rectus femoris

Vastus lateralis

Vastus medialis

Fibularis longus

Tibialis anterior

Extensor digitorum longus

Masseter

Orbicularis
oris

Sternocleidomastoid

Trapezius

Pectoralis minor

Serratus anterior

Rectus abdominis

Transversus
abdominis

Internal abdominal
oblique

External abdominal
oblique

Gracilis

Gastrocnemius

Soleus

(a)

Figure 10.4a   The Muscular System

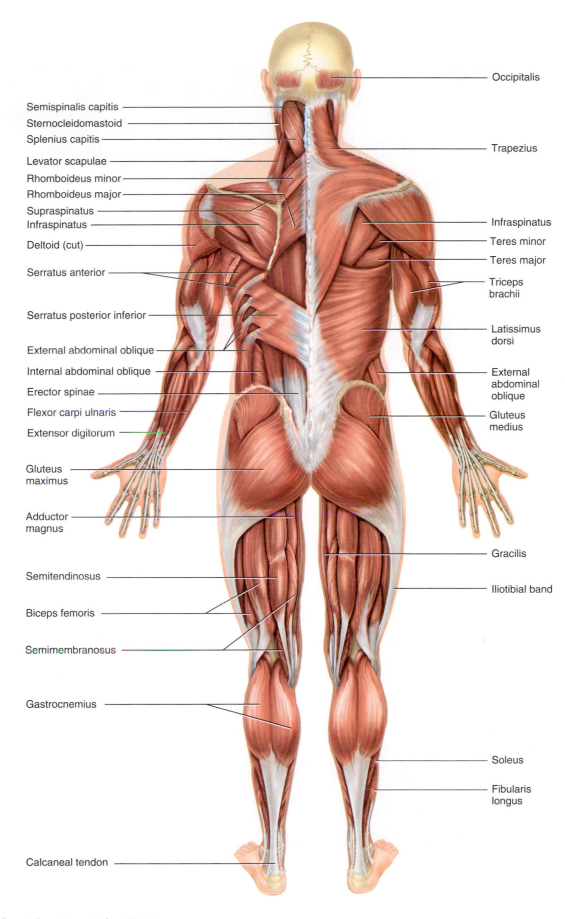

Occipitalis

Semispinalis capitis

Sternocleidomastoid

Splenius capitis

Trapezius

Levator scapulae

Rhomboideus minor

Rhomboideus major

Supraspinatus

Infraspinatus

Infraspinatus

Deltoid (cut)

Teres minor

Teres major

Serratus anterior

Triceps brachii

Serratus posterior inferior

Latissimus dorsi

External abdominal oblique

Internal abdominal oblique

External abdominal oblique

Erector spinae

Flexor carpi ulnaris

Gluteus medius

Extensor digitorum

Gluteus maximus

Adductor magnus

Gracilis

Iliotibial band

Semitendinosus

Biceps femoris

Semimembranosus

Gastrocnemius

Soleus

Fibularis longus

Calcaneal tendon

Figure 10.4b   The Muscular System

Frontalis

Procerus

Orbicularis
oculi

Nasalis

Levator labii
superioris

Zygomaticus
major

Orbicularis
oris

Parotid salivary
gland

Masseter

Depressor
labii inferioris

Depressor
anguli oris

Platysma

Sternocleidomastoid

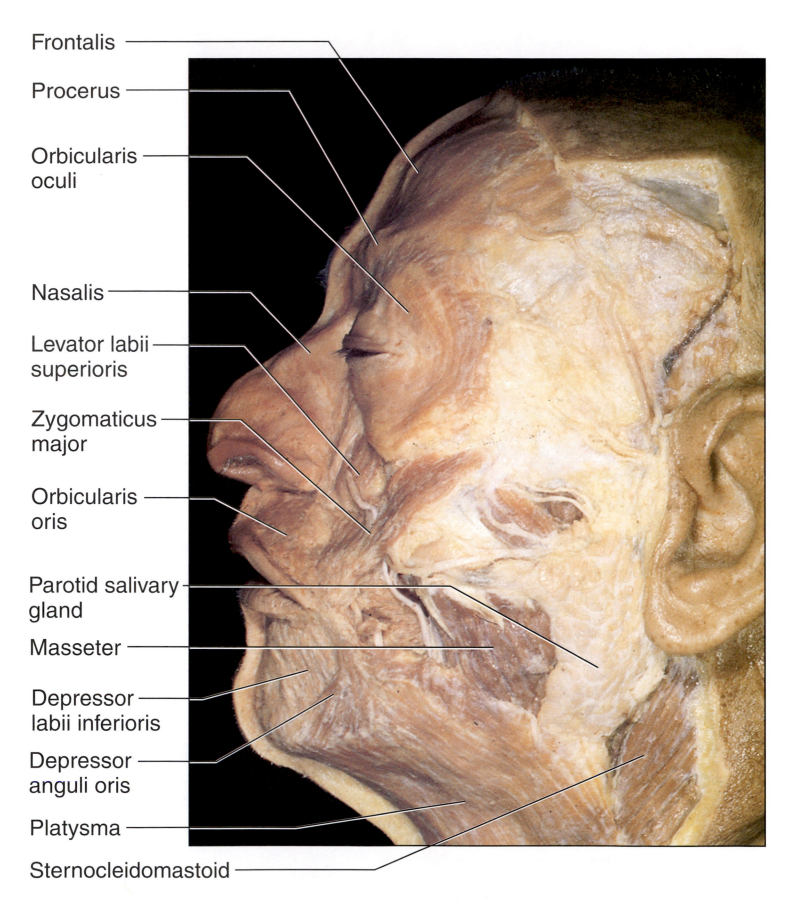

Figure 10.5 Some Muscles of Facial Expression in the Cadaver

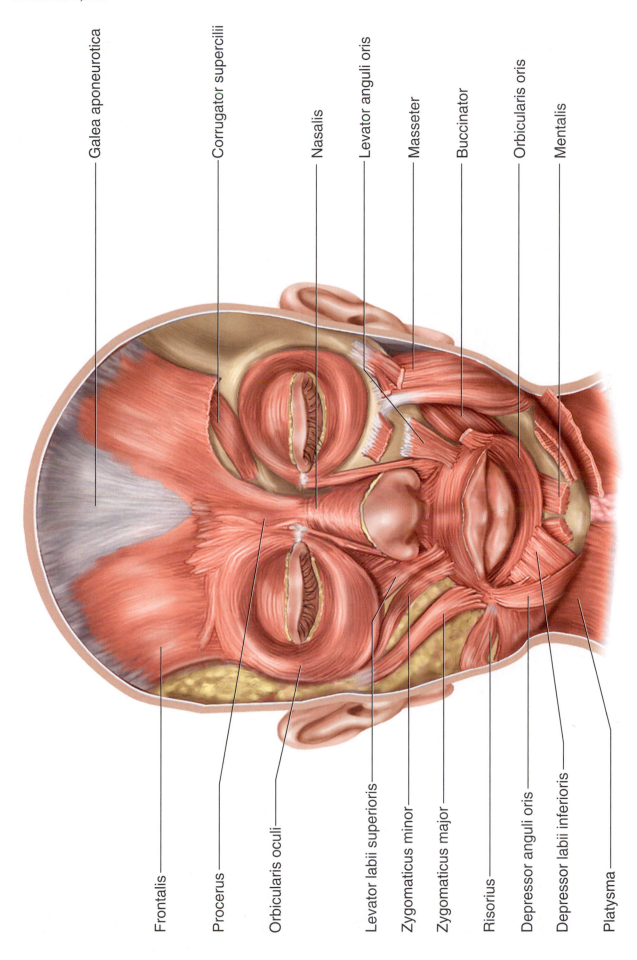

Galea aponeurotica

Corrugator supercilii

Nasalis

Levator anguli oris

Masseter

Buccinator

Orbicularis oris

Mentalis

Frontalis

Procerus

Orbicularis oculi

Levator labii superioris

Zygomaticus minor

Zygomaticus major

Risorius

Depressor anguli oris

Depressor labii inferioris

Platysma

Figure 10.7 top The Muscles of Facial Expression

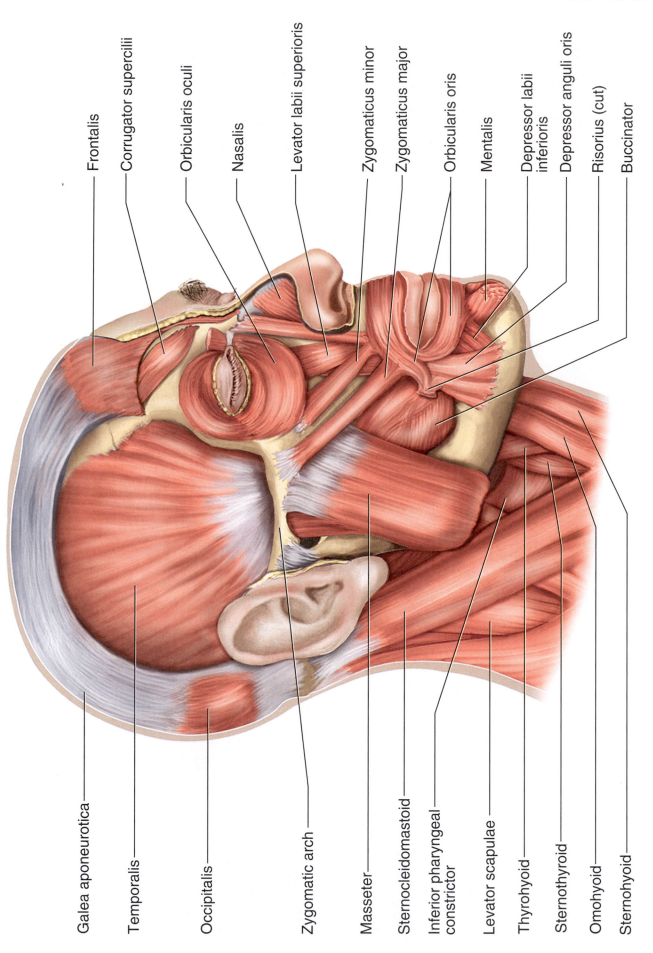

Frontalis

Corrugator supercilii

Orbicularis oculi

Nasalis

Levator labii superioris

Zygomaticus minor

Zygomaticus major

Orbicularis oris

Mentalis

Depressor labii inferioris

Depressor anguli oris

Risorius (cut)

Buccinator

Galea aponeurotica

Temporalis

Occipitalis

Zygomatic arch

Masseter

Sternocleidomastoid

Inferior pharyngeal constrictor

Levator scapulae

Thyrohyoid

Sternothyroid

Omohyoid

Sternohyoid

Figure 10.7 bottom The Muscles of Facial Expression

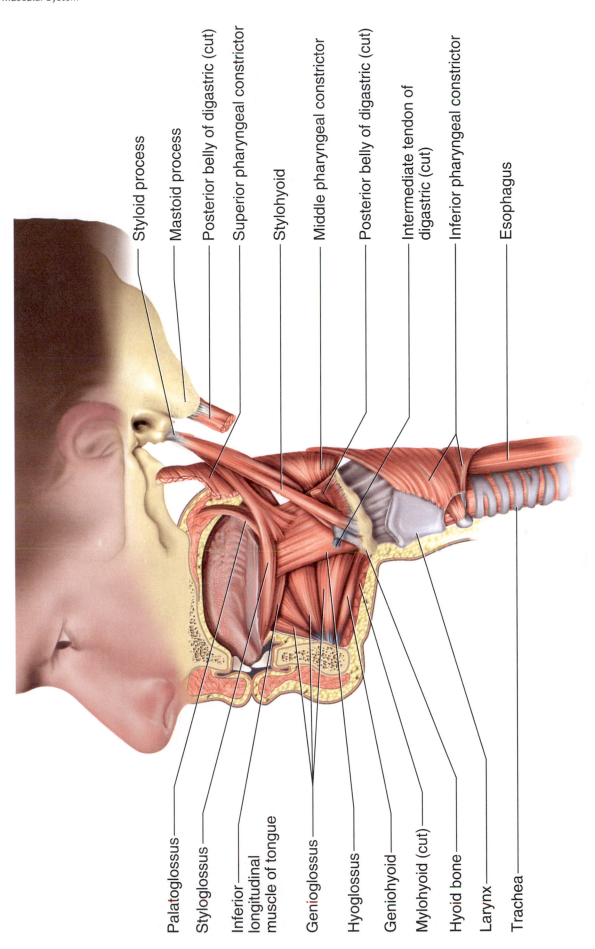

Styloid process

Mastoid process

Posterior belly of digastric (cut)

Superior pharyngeal constrictor

Stylohyoid

Middle pharyngeal constrictor

Posterior belly of digastric (cut)

Intermediate tendon of digastric (cut)

Inferior pharyngeal constrictor

Esophagus

Palatoglossus

Styloglossus

Inferior longitudinal muscle of tongue

Genioglossus

Hyoglossus

Geniohyoid

Mylohyoid (cut)

Hyoid bone

Larynx

Trachea

Figure 10.8   Muscles of the Tongue and Pharynx

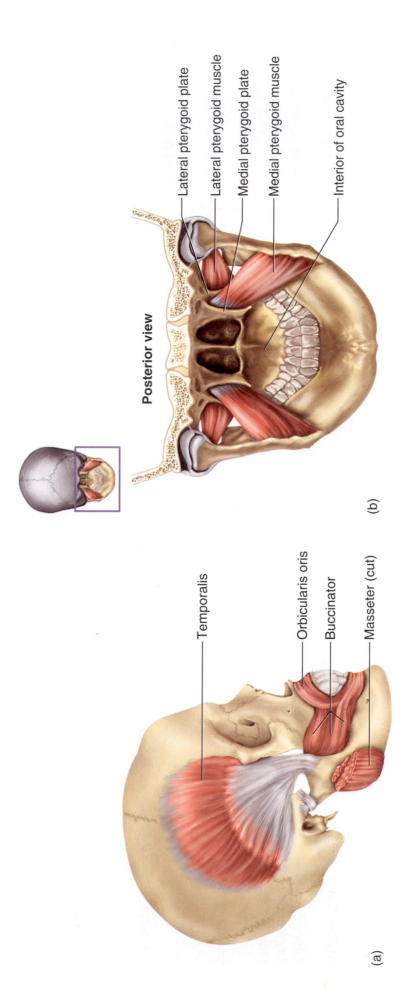

Lateral pterygoid plate

Lateral pterygoid muscle

Medial pterygoid plate

Medial pterygoid muscle

Interior of oral cavity

Posterior view

(b)

Temporalis

Orbicularis oris

Buccinator

Masseter (cut)

(a)

Figure 10.9   Muscles of Chewing

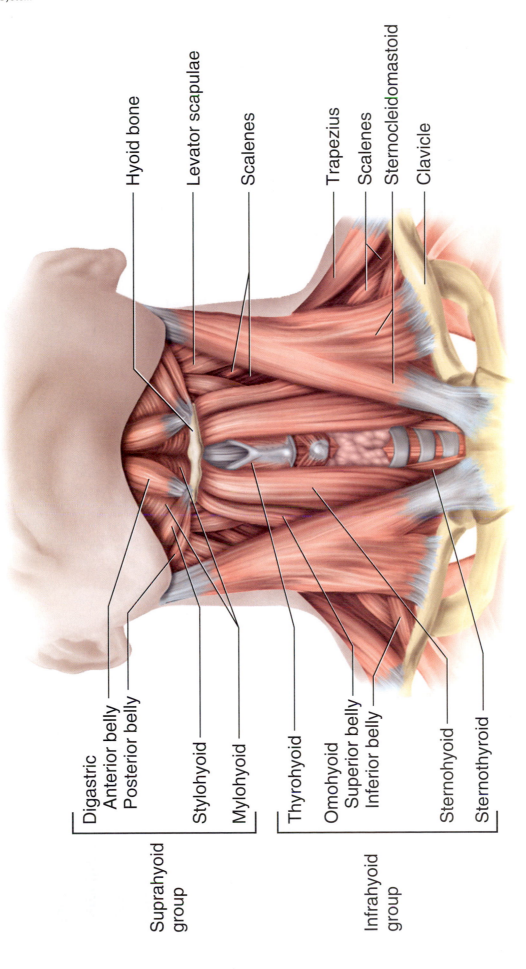

Hyoid bone

Levator scapulae

Scalenes

Trapezius

Scalenes

Sternocleidomastoid

Clavicle

Digastric
Anterior belly
Posterior belly

Suprahyoid group

Stylohyoid

Mylohyoid

Thyrohyoid

Omohyoid
Superior belly
Inferior belly

Infrahyoid group

Sternohyoid

Sternothyroid

Figure 10.10a   Muscles of the Neck—Anterior View

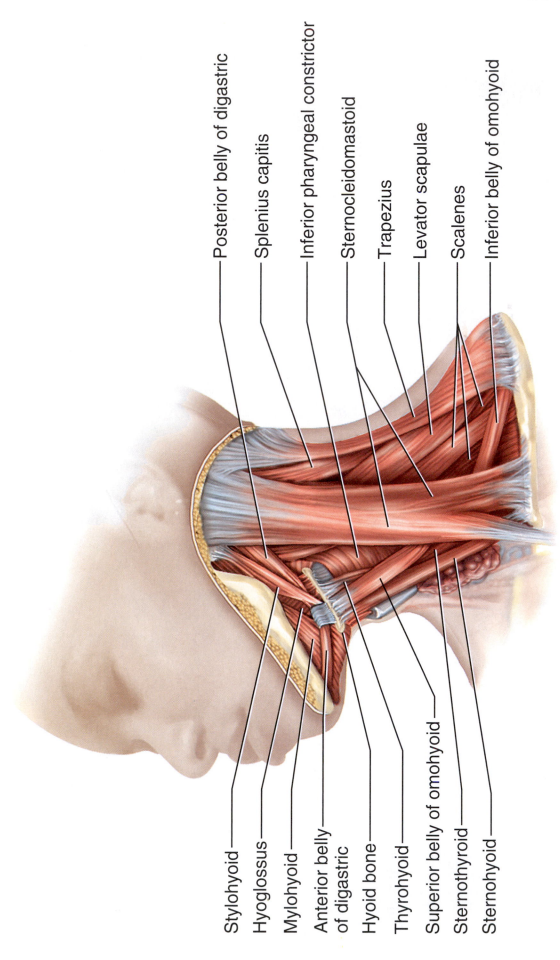

Posterior belly of digastric

Splenius capitis

Inferior pharyngeal constrictor

Sternocleidomastoid

Trapezius

Levator scapulae

Scalenes

Inferior belly of omohyoid

Stylohyoid

Hyoglossus

Mylohyoid

Anterior belly of digastric

Hyoid bone

Thyrohyoid

Superior belly of omohyoid

Sternothyroid

Sternohyoid

Figure 10.10b   Muscles of the Neck—Left Lateral View

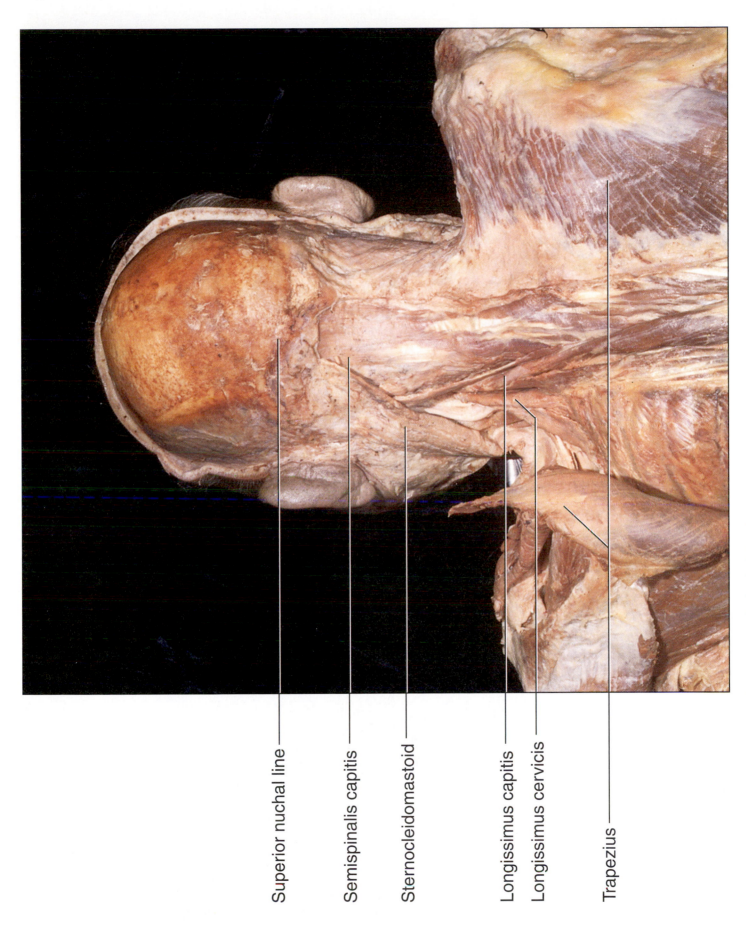

Superior nuchal line

Semispinalis capitis

Sternocleidomastoid

Longissimus capitis

Longissimus cervicis

Trapezius

Figure 10.12   Muscles of the Shoulder and Nuchal Regions

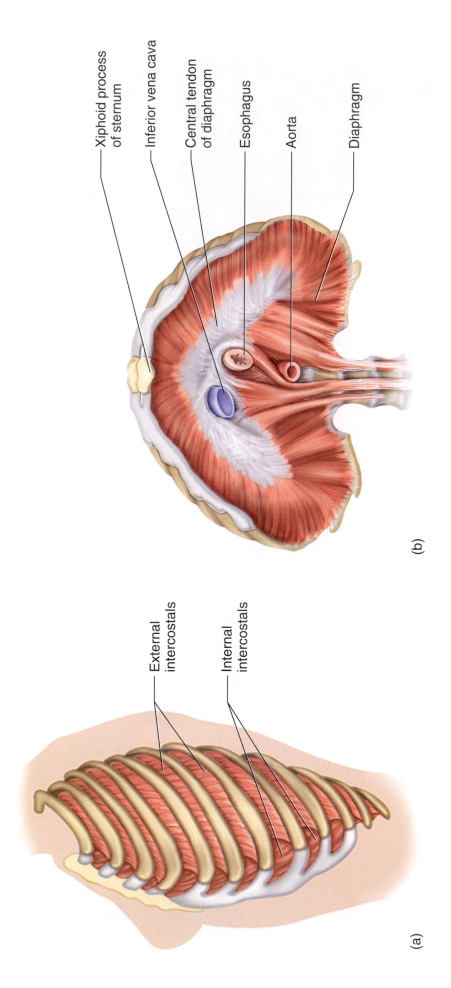

Xiphoid process
of sternum

Inferior vena cava

Central tendon
of diaphragm

Esophagus

Aorta

Diaphragm

(b)

External
intercostals

Internal
intercostals

(a)

Figure 10.13   Muscles of Respiration

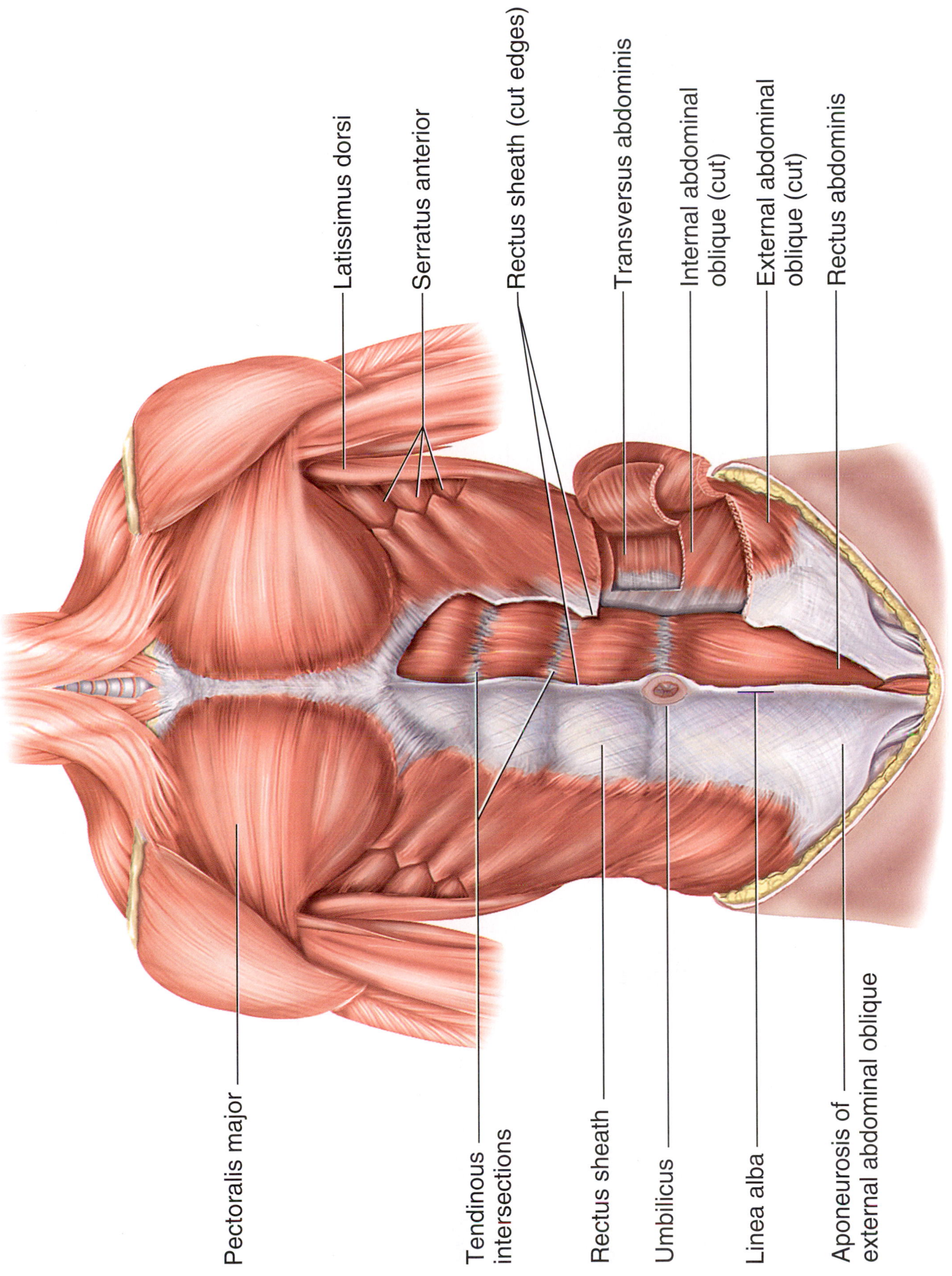

Latissimus dorsi

Serratus anterior

Rectus sheath (cut edges)

Transversus abdominis

Internal abdominal oblique (cut)

External abdominal oblique (cut)

Rectus abdominis

Pectoralis major

Tendinous intersections

Rectus sheath

Umbilicus

Linea alba

Aponeurosis of external abdominal oblique

Figure 10.15a Thoracic and Abdominal Muscles—Superficial Muscles

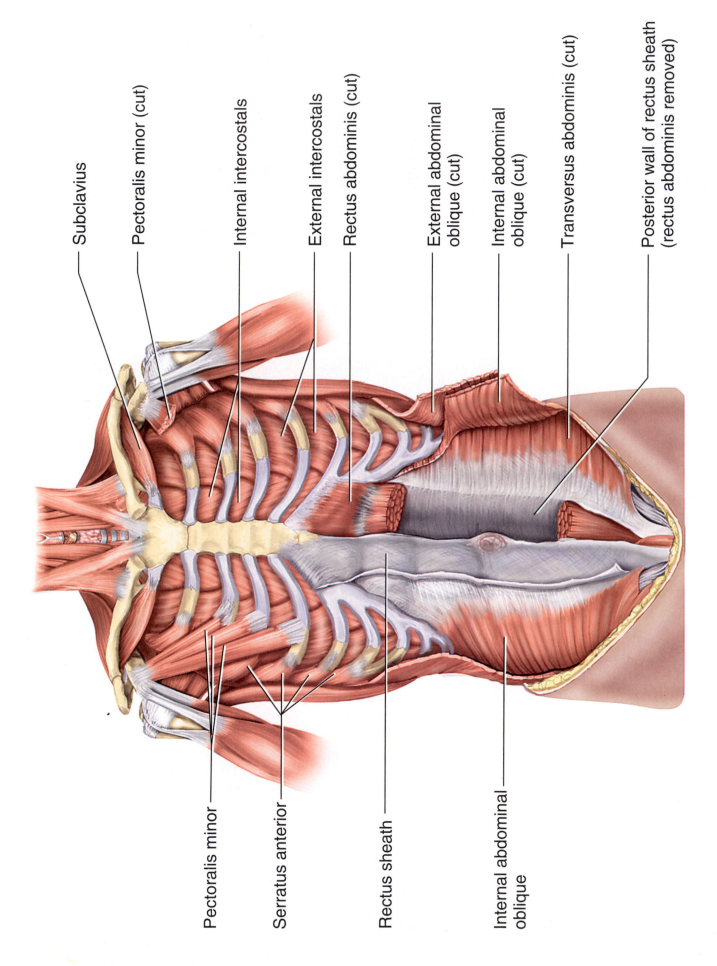

Subclavius

Pectoralis minor (cut)

Internal intercostals

External intercostals

Rectus abdominis (cut)

External abdominal oblique (cut)

Internal abdominal oblique (cut)

Transversus abdominis (cut)

Posterior wall of rectus sheath (rectus abdominis removed)

Pectoralis minor

Serratus anterior

Rectus sheath

Internal abdominal oblique

Figure 10.15b   Thoracic and Abdominal Muscles—Deep Muscles

Superficial muscles | **Deep muscles**

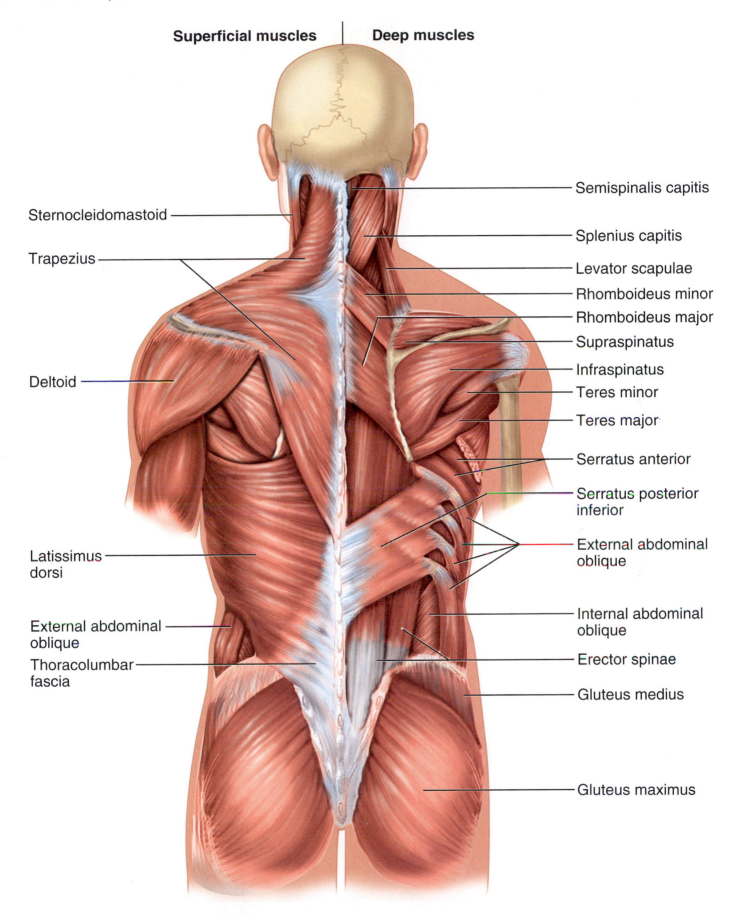

Sternocleidomastoid

Trapezius

Deltoid

Latissimus dorsi

External abdominal oblique

Thoracolumbar fascia

Semispinalis capitis

Splenius capitis

Levator scapulae

Rhomboideus minor

Rhomboideus major

Supraspinatus

Infraspinatus

Teres minor

Teres major

Serratus anterior

Serratus posterior inferior

External abdominal oblique

Internal abdominal oblique

Erector spinae

Gluteus medius

Gluteus maximus

Figure 10.17   Neck, Back, and Gluteal Muscles

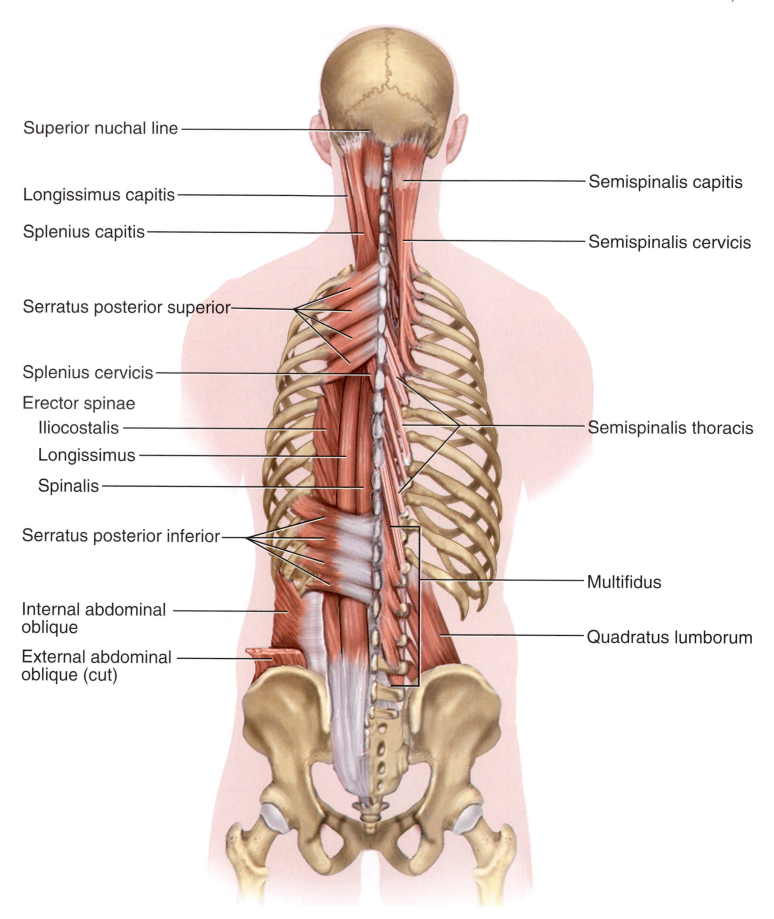

Superior nuchal line

Longissimus capitis

Splenius capitis

Serratus posterior superior

Splenius cervicis

Erector spinae
  Iliocostalis
  Longissimus
  Spinalis

Serratus posterior inferior

Internal abdominal oblique

External abdominal oblique (cut)

Semispinalis capitis

Semispinalis cervicis

Semispinalis thoracis

Multifidus

Quadratus lumborum

Figure 10.18   **Muscles Acting on the Vertebral Column**

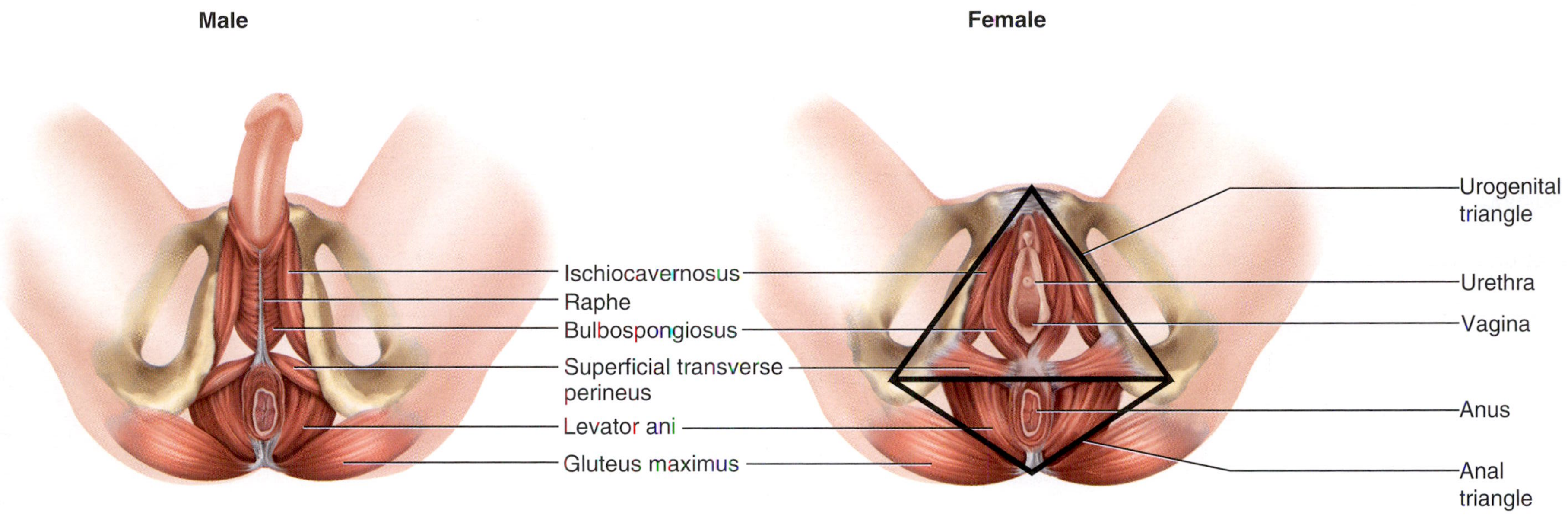

Urogenital triangle
Urethra
Vagina
Anus
Anal triangle

Female

Ischiocavernosus
Raphe
Bulbospongiosus
Superficial transverse perineus
Levator ani
Gluteus maximus

Male

Figure 10.20a and b   Muscles of the Pelvic Floor

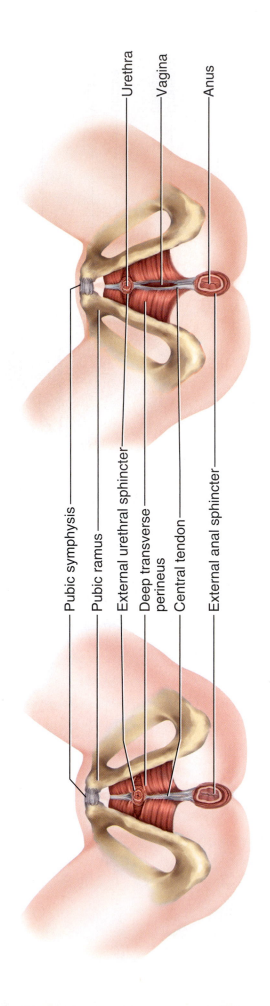

Figure 10.20c and d   Muscles of the Pelvic Floor

Urethra

Vagina

Anus

Pubic symphysis

Pubic ramus

External urethral sphincter

Deep transverse perineus

Central tendon

External anal sphincter

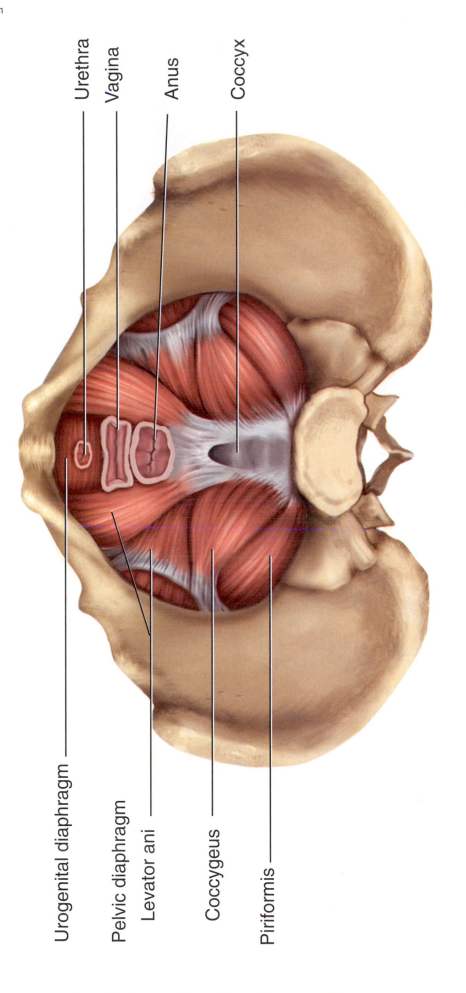

Urethra

Vagina

Anus

Coccyx

Urogenital diaphragm

Pelvic diaphragm

Levator ani

Coccygeus

Piriformis

Figure 10.20e The Pelvic Diaphragm, the Deepest Layer, Superior View

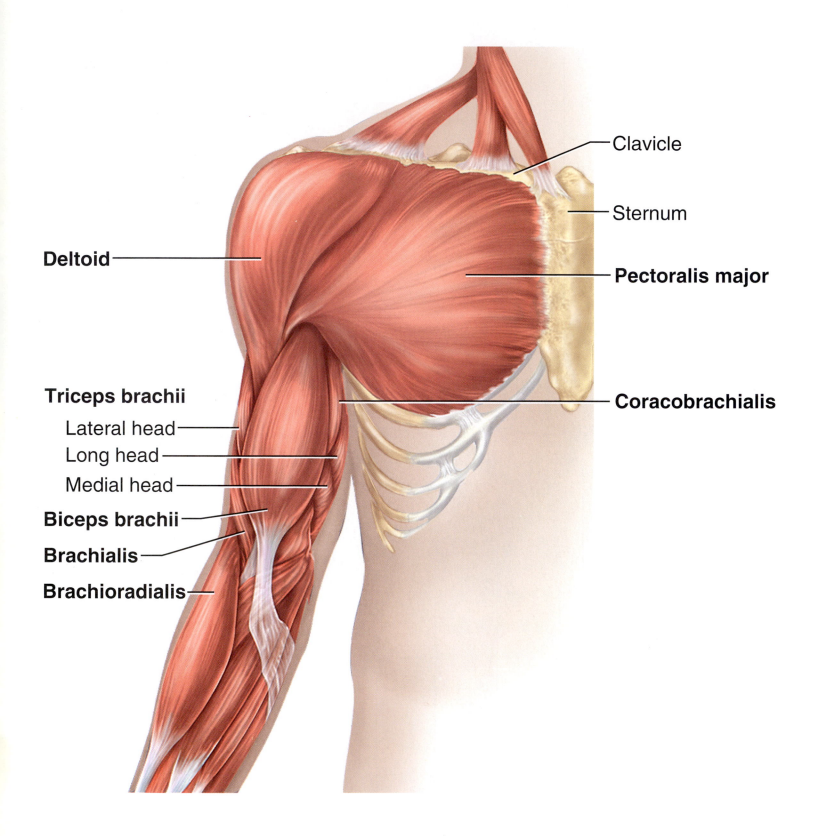

Deltoid

Triceps brachii

Lateral head

Long head

Medial head

Biceps brachii

Brachialis

Brachioradialis

Clavicle

Sternum

Pectoralis major

Coracobrachialis

Figure 10.22a   Pectoral and Brachial Muscles—Anterior View

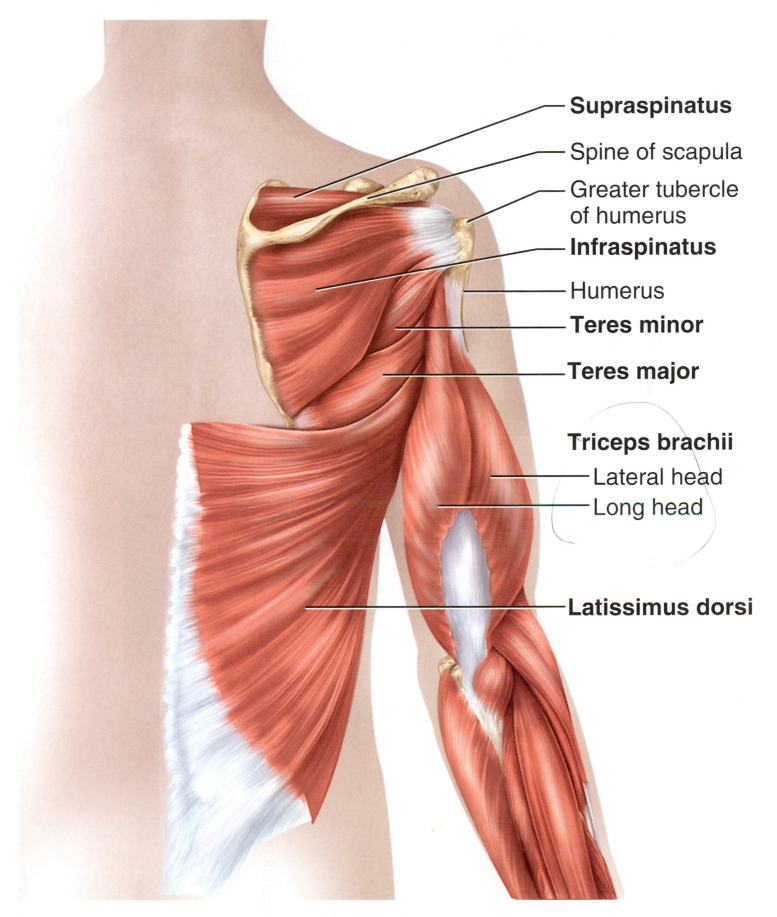

Supraspinatus

Spine of scapula

Greater tubercle of humerus

Infraspinatus

Humerus

Teres minor

Teres major

Triceps brachii

Lateral head

Long head

Latissimus dorsi

Figure 10.22b Pectoral and Brachial Muscles—Posterior View

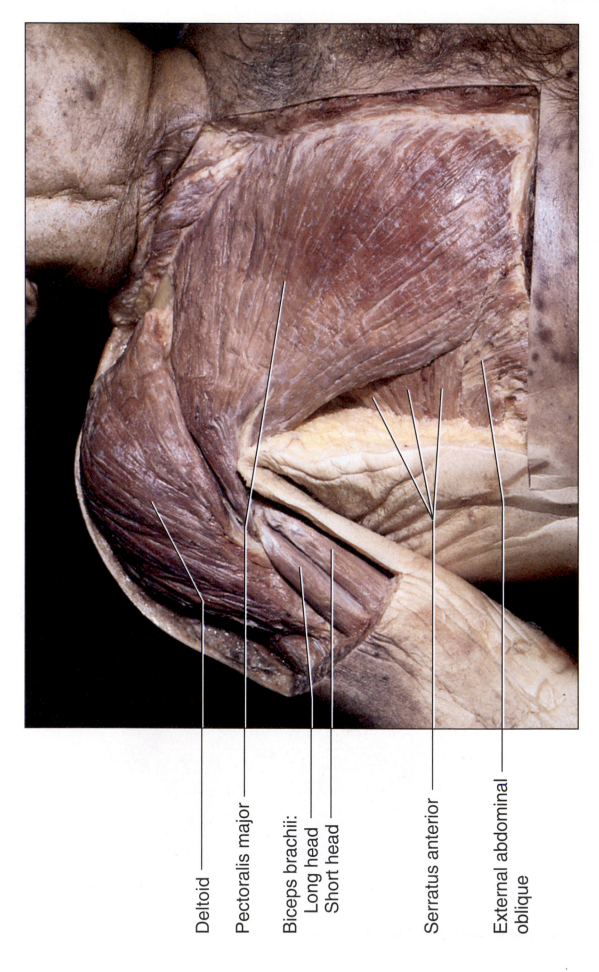

Deltoid

Pectoralis major

Biceps brachii:
Long head
Short head

Serratus anterior

External abdominal
oblique

Figure 10.23a   Muscles of the Chest and Brachial Region of the Cadaver—Anterior View

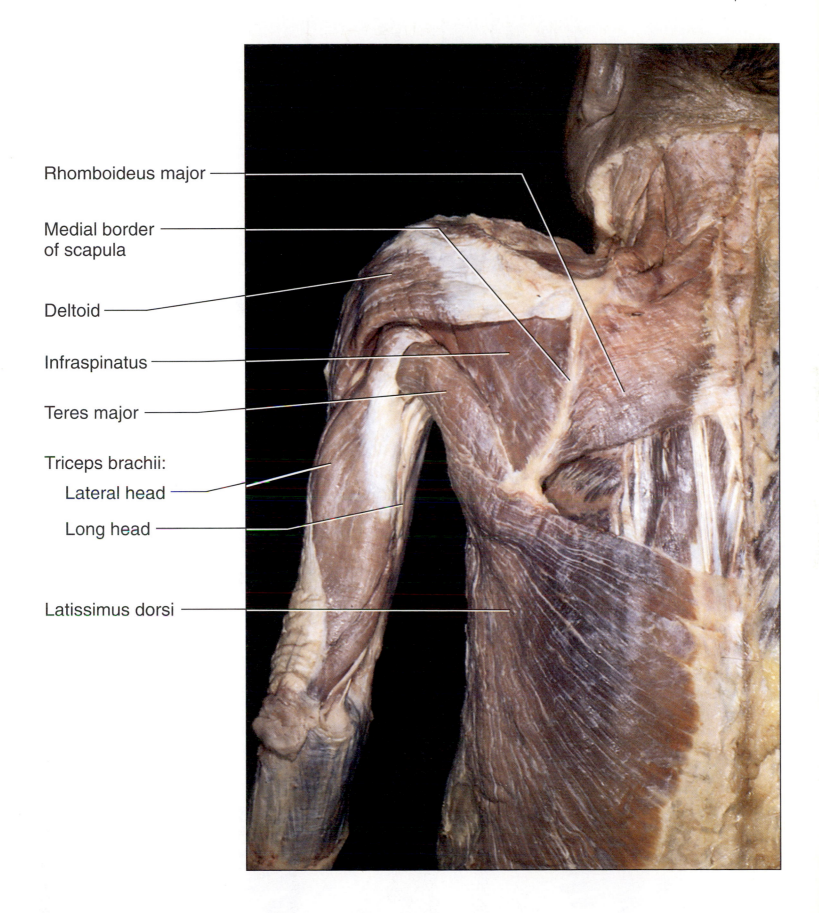

Rhomboideus major

Medial border
of scapula

Deltoid

Infraspinatus

Teres major

Triceps brachii:

   Lateral head

   Long head

Latissimus dorsi

Figure 10.23b   **Muscles of the Chest and Brachial Region of the Cadaver—Posterior View**

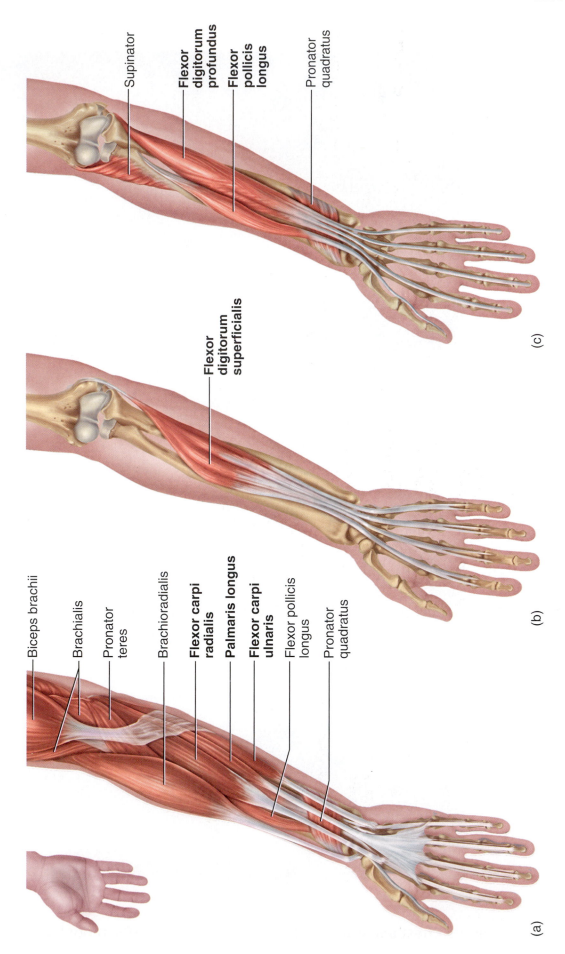

Supinator

Flexor digitorum profundus

Flexor pollicis longus

Pronator quadratus

Flexor digitorum superficialis

(c)

Biceps brachii

Brachialis

Pronator teres

Brachioradialis

Flexor carpi radialis

Palmaris longus

Flexor carpi ulnaris

Flexor pollicis longus

Pronator quadratus

(b)

(a)

Figure 10.28a-c   Muscles of the Forearm—Anterior View

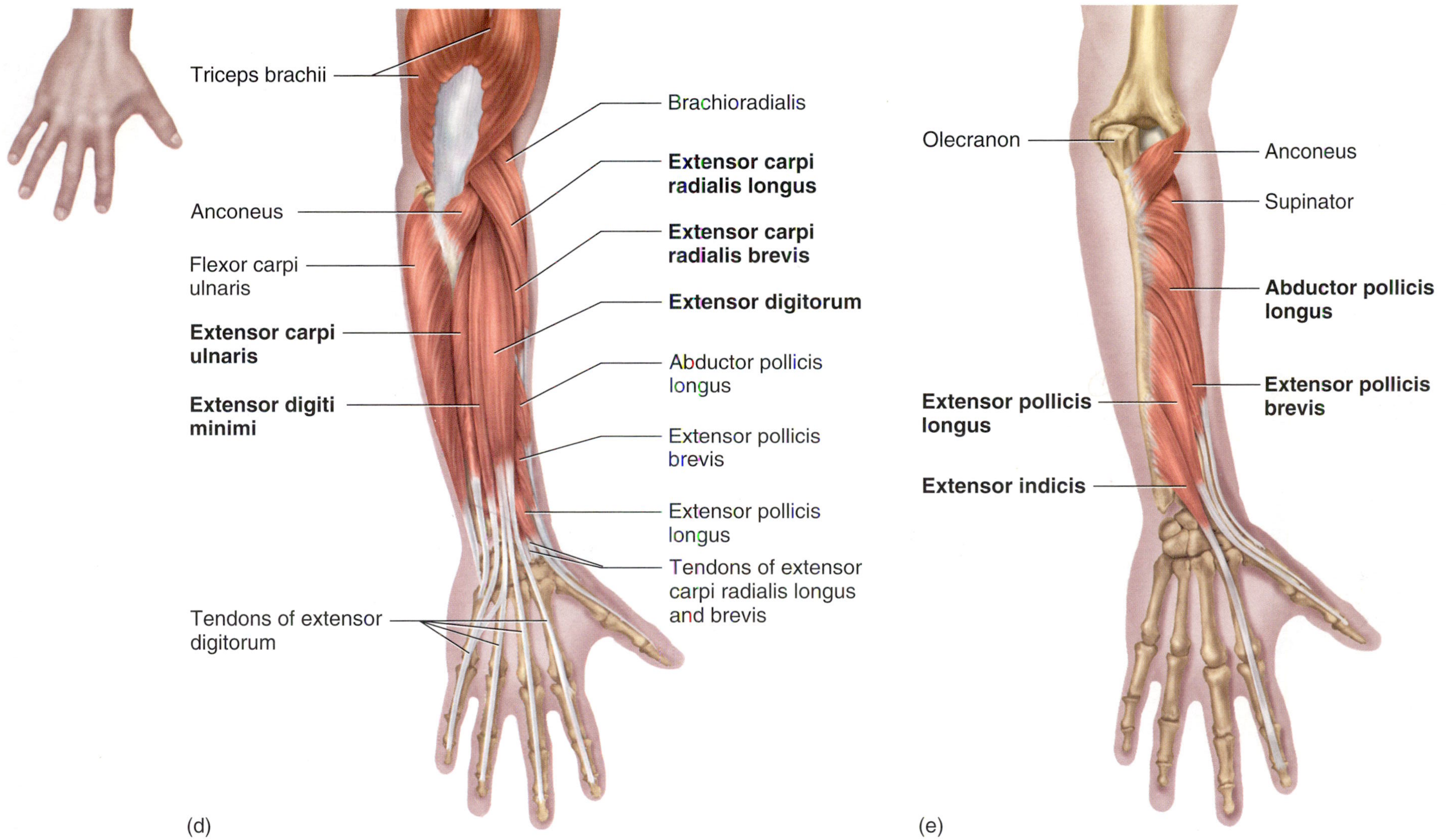

(e)

Anconeus

Supinator

Abductor pollicis longus

Extensor pollicis brevis

Olecranon

Extensor pollicis longus

Extensor indicis

(d)

Brachioradialis

Extensor carpi radialis longus

Extensor carpi radialis brevis

Extensor digitorum

Abductor pollicis longus

Extensor pollicis brevis

Extensor pollicis longus

Tendons of extensor carpi radialis longus and brevis

Triceps brachii

Anconeus

Flexor carpi ulnaris

Extensor carpi ulnaris

Extensor digiti minimi

Tendons of extensor digitorum

Figure 10.28d-e   Muscles of the Forearm—Posterior View

Palmar aspect, superficial

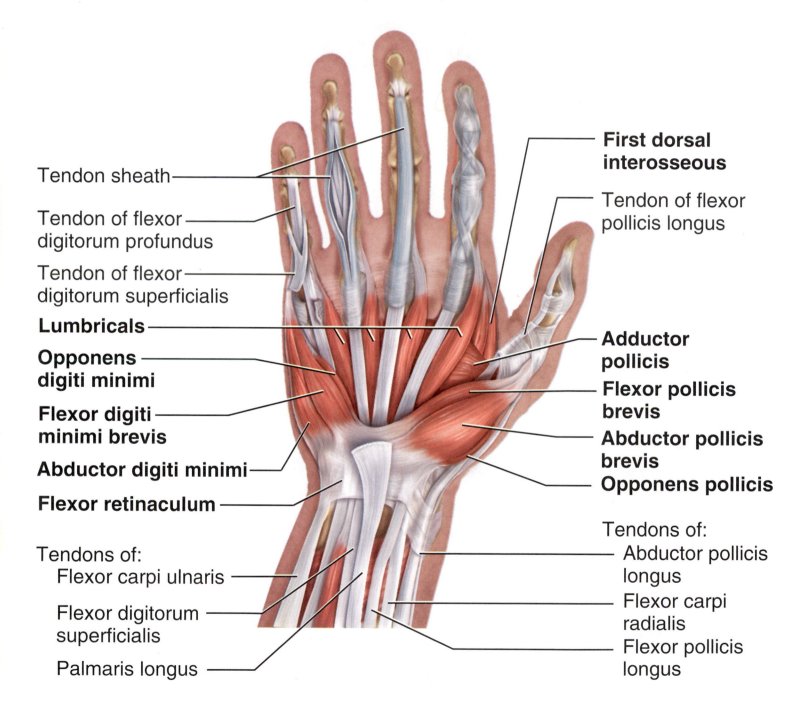

Tendon sheath

Tendon of flexor
digitorum profundus

Tendon of flexor
digitorum superficialis

Lumbricals

**Opponens
digiti minimi**

**Flexor digiti
minimi brevis**

Abductor digiti minimi

Flexor retinaculum

Tendons of:
　Flexor carpi ulnaris

　Flexor digitorum
　superficialis

　Palmaris longus

**First dorsal
interosseous**

Tendon of flexor
pollicis longus

**Adductor
pollicis**

**Flexor pollicis
brevis**

**Abductor pollicis
brevis**

Opponens pollicis

Tendons of:
　Abductor pollicis
　longus

　Flexor carpi
　radialis

　Flexor pollicis
　longus

Figure 10.29a　Intrinsic Muscles of the Hand—Superficial Muscles

Palmar aspect, deep

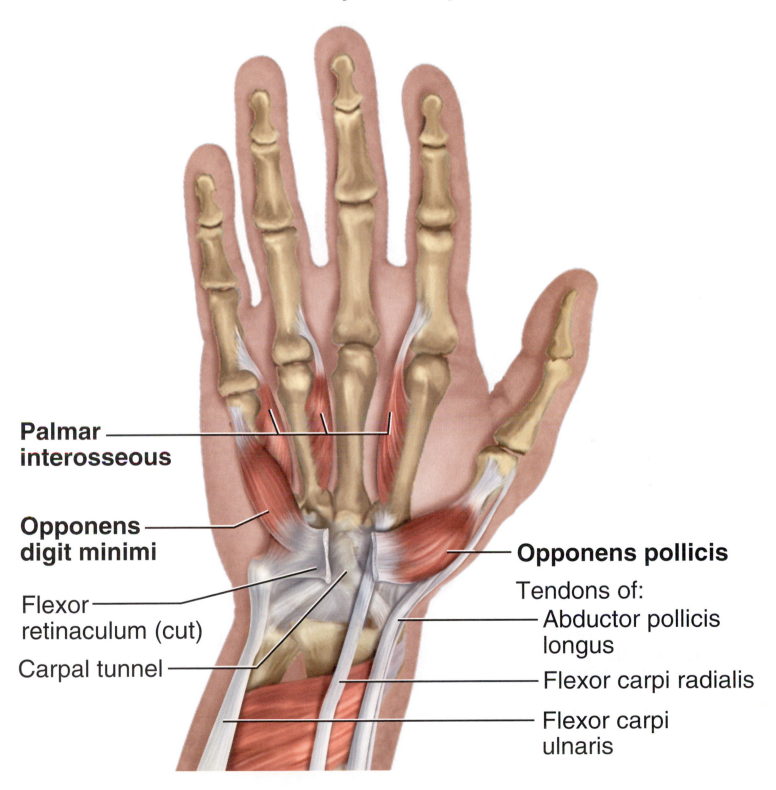

Palmar interosseous

Opponens digit minimi

Flexor retinaculum (cut)

Carpal tunnel

Opponens pollicis

Tendons of:

Abductor pollicis longus

Flexor carpi radialis

Flexor carpi ulnaris

Figure 10.29b   Intrinsic Muscles of the Hand—Deep Muscles, Anterior View

Dorsal aspect

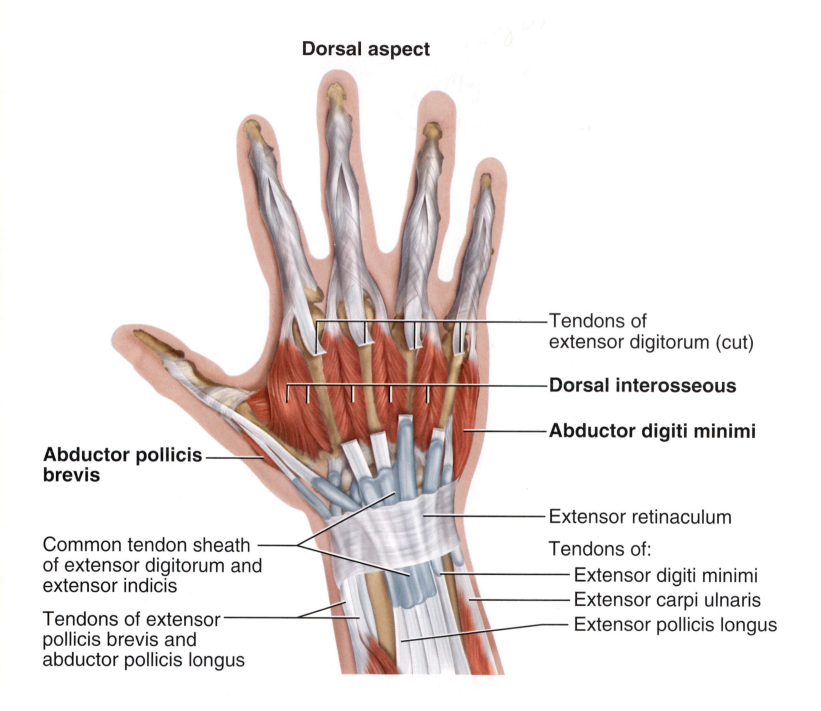

Tendons of
extensor digitorum (cut)

Dorsal interosseous

Abductor digiti minimi

**Abductor pollicis
brevis**

Extensor retinaculum

Tendons of:

Common tendon sheath
of extensor digitorum and
extensor indicis

Extensor digiti minimi

Extensor carpi ulnaris

Tendons of extensor
pollicis brevis and
abductor pollicis longus

Extensor pollicis longus

Figure 10.29c   Intrinsic Muscles of the Hand—Superficial Muscles

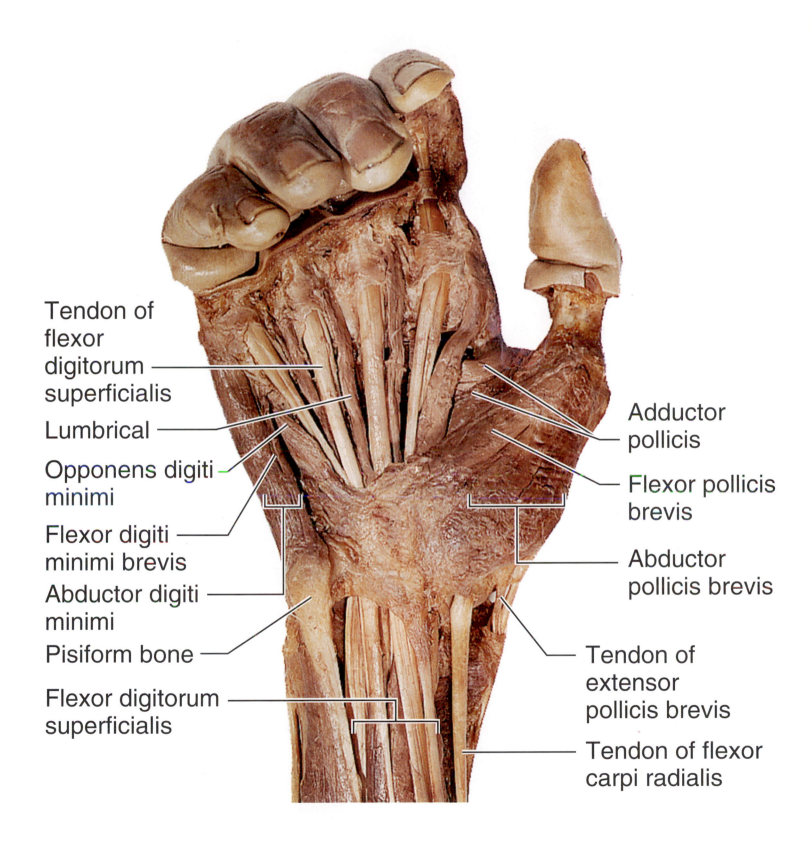

Tendon of
flexor
digitorum
superficialis

Lumbrical

Opponens digiti
minimi

Flexor digiti
minimi brevis

Abductor digiti
minimi

Pisiform bone

Flexor digitorum
superficialis

Adductor
pollicis

Flexor pollicis
brevis

Abductor
pollicis brevis

Tendon of
extensor
pollicis brevis

Tendon of flexor
carpi radialis

Figure 10.29d Intrinsic Muscles of the Hand—Anterior View of Cadaver Hand

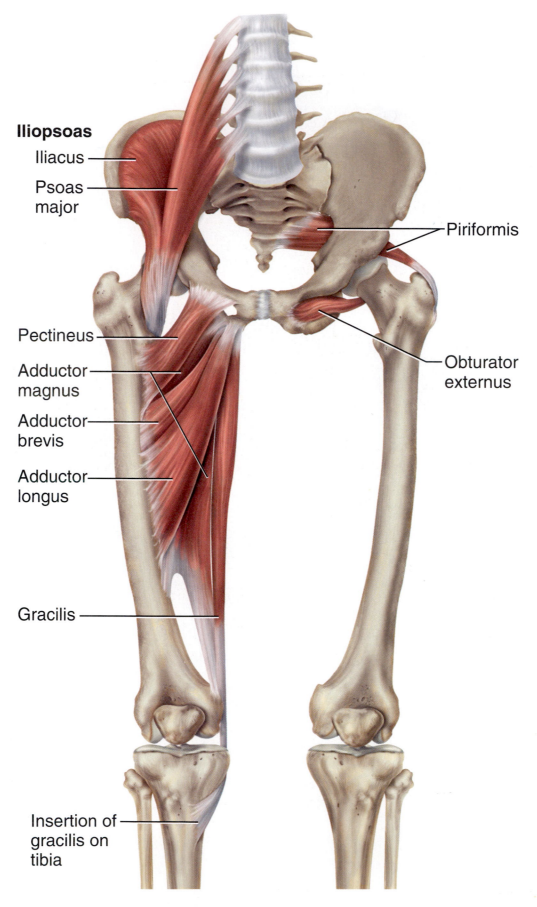

Iliopsoas
Iliacus
Psoas major
Pectineus
Adductor magnus
Adductor brevis
Adductor longus
Gracilis
Insertion of gracilis on tibia
Piriformis
Obturator externus

Figure 10.30   Muscles Acting on the Hip and Femur

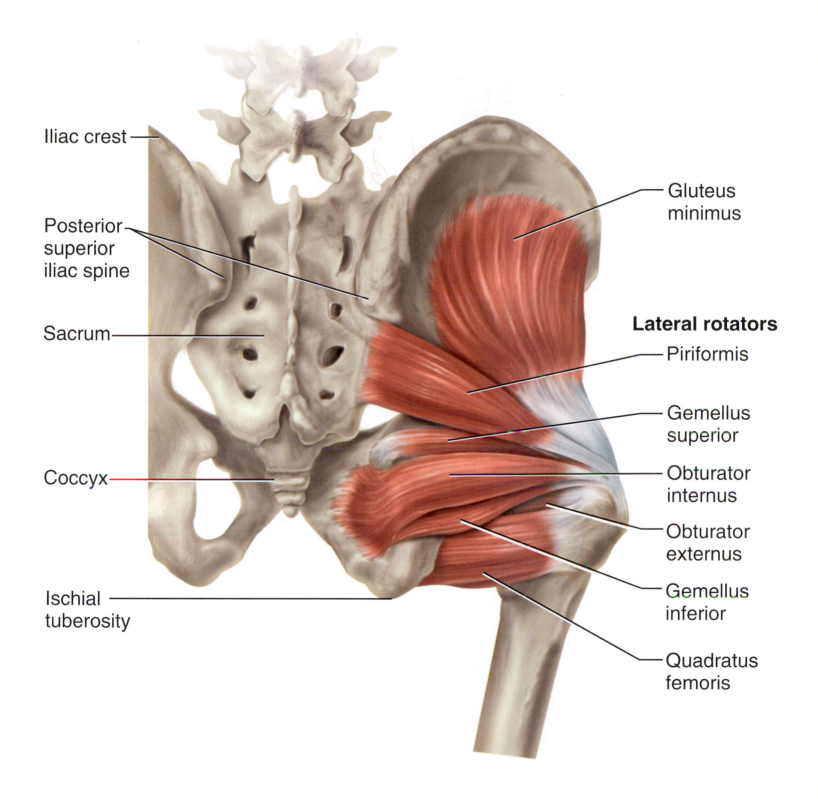

Iliac crest

Posterior superior iliac spine

Sacrum

Coccyx

Ischial tuberosity

Gluteus minimus

Lateral rotators

Piriformis

Gemellus superior

Obturator internus

Obturator externus

Gemellus inferior

Quadratus femoris

Figure 10.31 Deep Gluteal Muscles

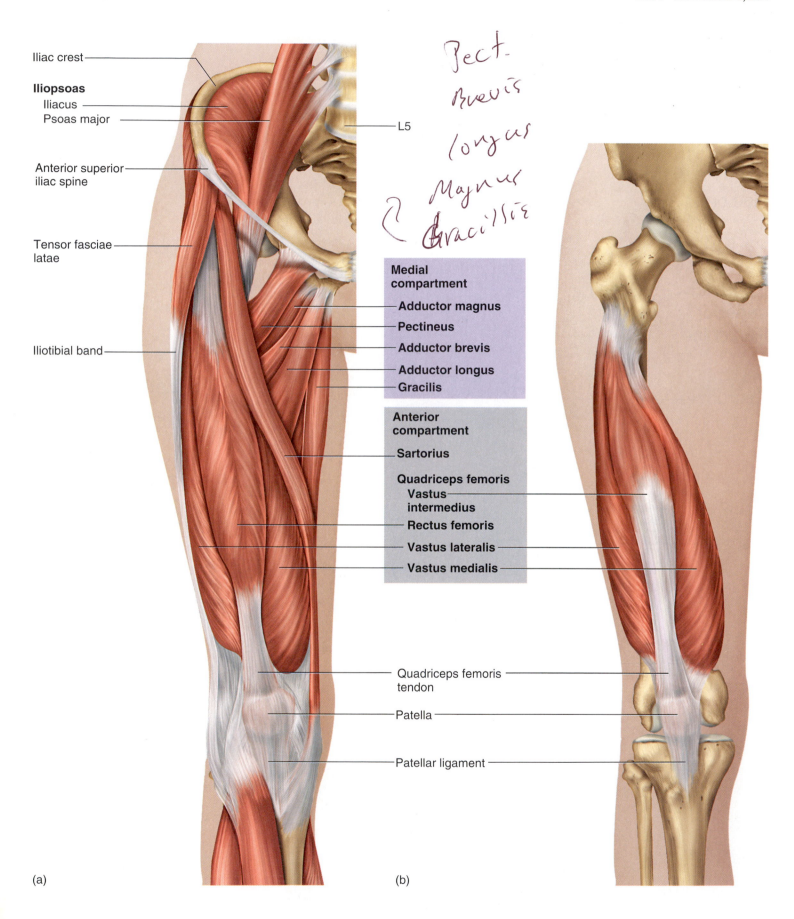

Iliac crest

Iliopsoas
  Iliacus
  Psoas major

Anterior superior
iliac spine

Tensor fasciae
latae

Iliotibial band

L5

Pect.
Brevis
longus
Magnus
Gracillis

Medial
compartment
  Adductor magnus
  Pectineus
  Adductor brevis
  Adductor longus
  Gracilis

Anterior
compartment
  Sartorius

  Quadriceps femoris
    **Vastus
    intermedius**
    Rectus femoris
    Vastus lateralis
    Vastus medialis

Quadriceps femoris
tendon

Patella

Patellar ligament

(a)   (b)

Figure 10.32   Muscles of the Thigh

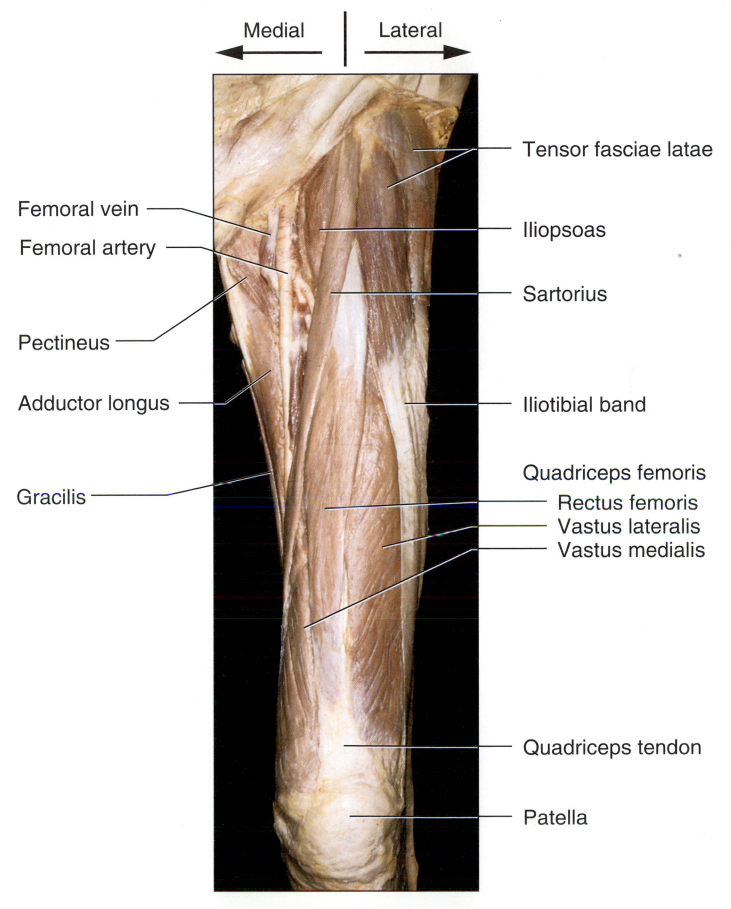

Medial ⟵   Lateral ⟶

Femoral vein

Femoral artery

Pectineus

Adductor longus

Gracilis

Tensor fasciae latae

Iliopsoas

Sartorius

Iliotibial band

Quadriceps femoris
Rectus femoris
Vastus lateralis
Vastus medialis

Quadriceps tendon

Patella

Figure 10.33 **Superficial Anterior Muscles of the Thigh of the Cadaver**

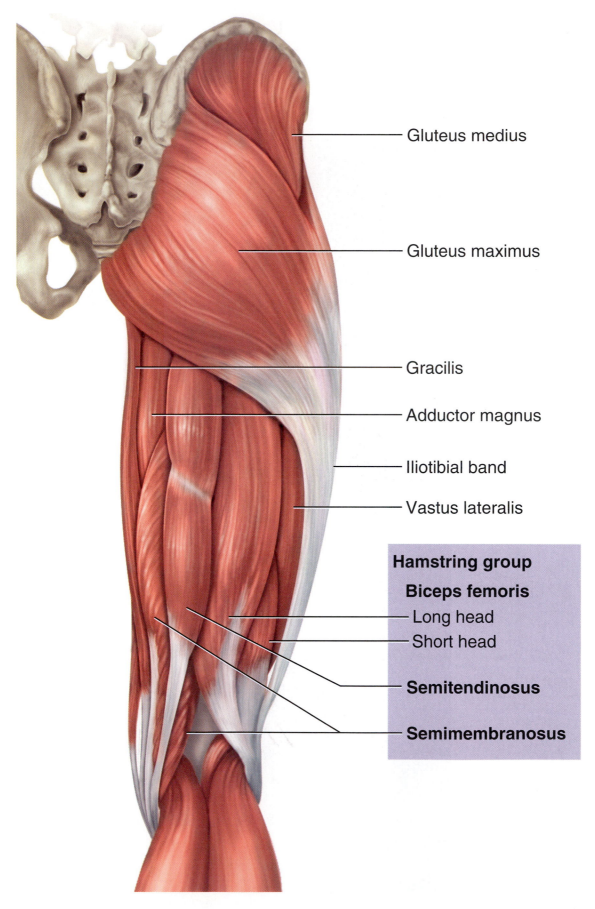

Gluteus medius

Gluteus maximus

Gracilis

Adductor magnus

Iliotibial band

Vastus lateralis

Hamstring group

Biceps femoris

Long head

Short head

Semitendinosus

Semimembranosus

Figure 10.34   Gluteal and Thigh Muscles

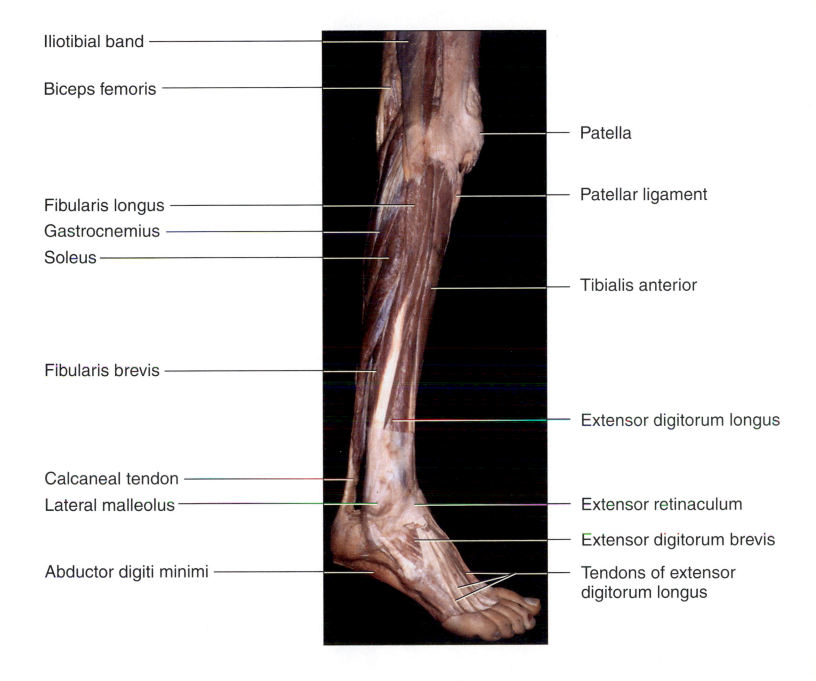

Iliotibial band

Biceps femoris

Fibularis longus

Gastrocnemius

Soleus

Fibularis brevis

Calcaneal tendon

Lateral malleolus

Abductor digiti minimi

Patella

Patellar ligament

Tibialis anterior

Extensor digitorum longus

Extensor retinaculum

Extensor digitorum brevis

Tendons of extensor digitorum longus

Figure 10.35 **Superficial Muscles of the Leg of the Cadaver**

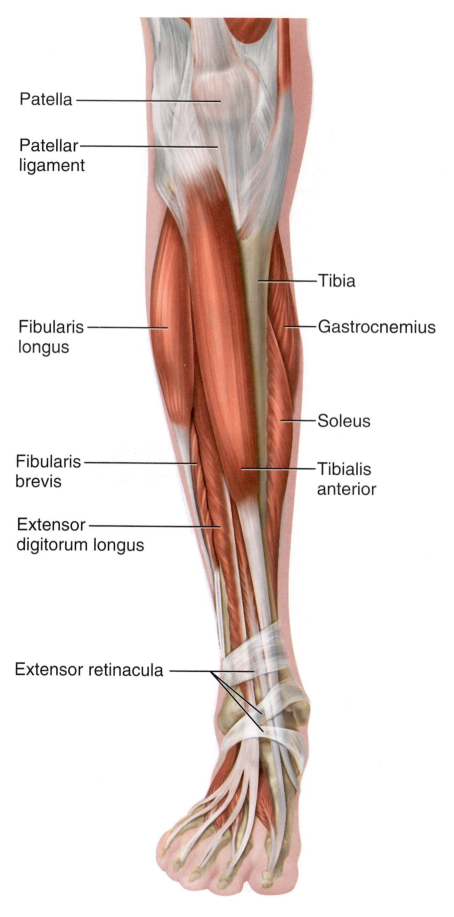

Patella

Patellar ligament

Tibia

Fibularis longus

Gastrocnemius

Soleus

Fibularis brevis

Tibialis anterior

Extensor digitorum longus

Extensor retinacula

Figure 10.36a **Muscles of the Leg**

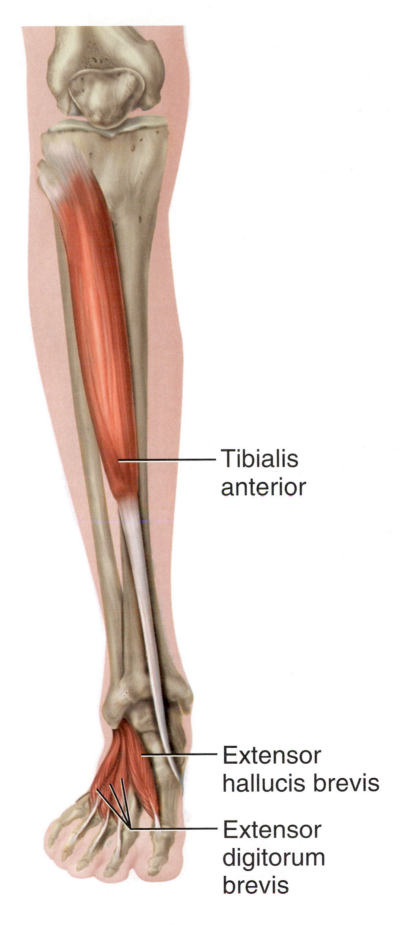

Tibialis anterior

Extensor hallucis brevis

Extensor digitorum brevis

Figure 10.36b   Muscles of the Leg

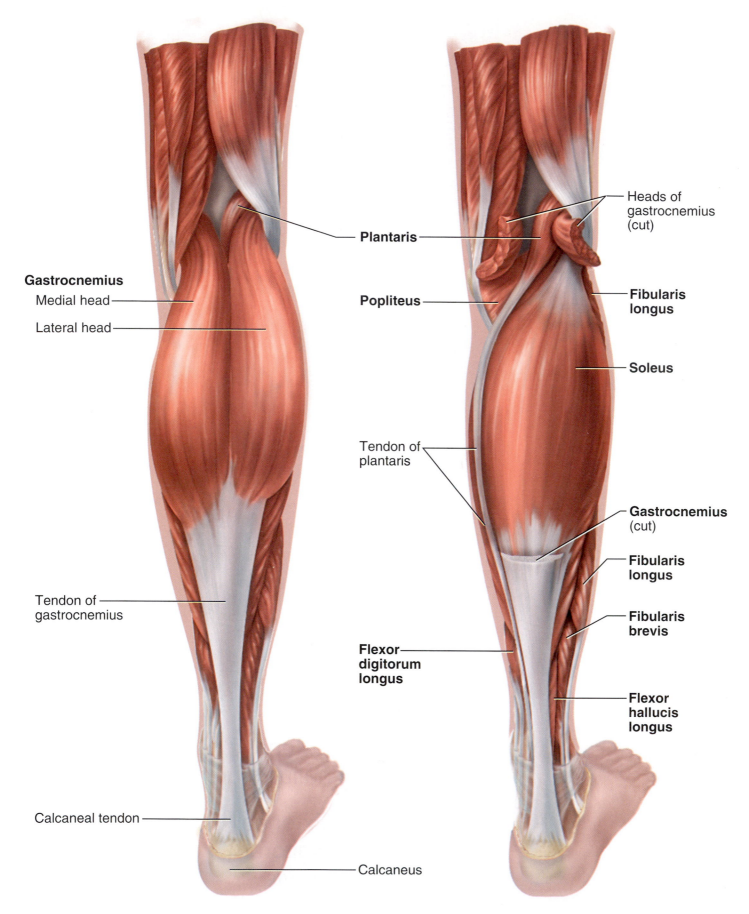

Figure 10.37   Superficial Muscles of the Leg, Posterior Compartment

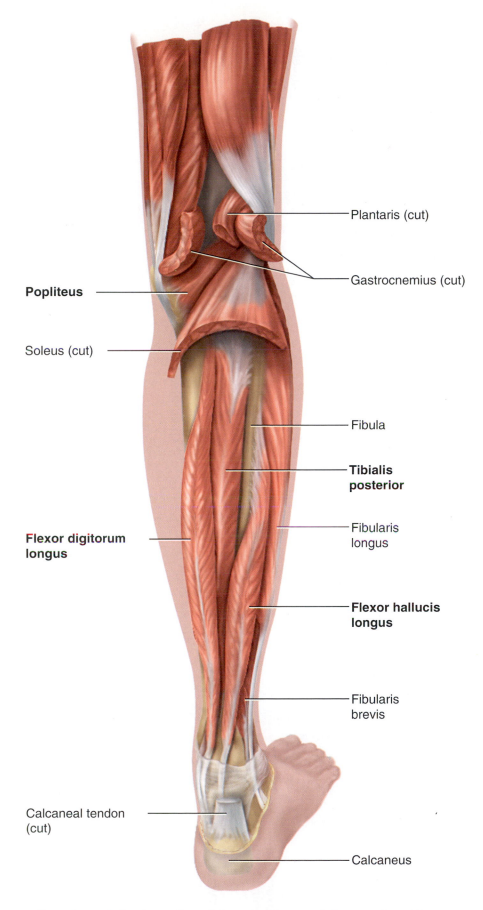

Plantaris (cut)

Gastrocnemius (cut)

Popliteus

Soleus (cut)

Fibula

Tibialis posterior

Fibularis longus

Flexor digitorum longus

Flexor hallucis longus

Fibularis brevis

Calcaneal tendon (cut)

Calcaneus

Figure 10.38a   Deep Muscles of the Leg, Posterior and Lateral Compartments

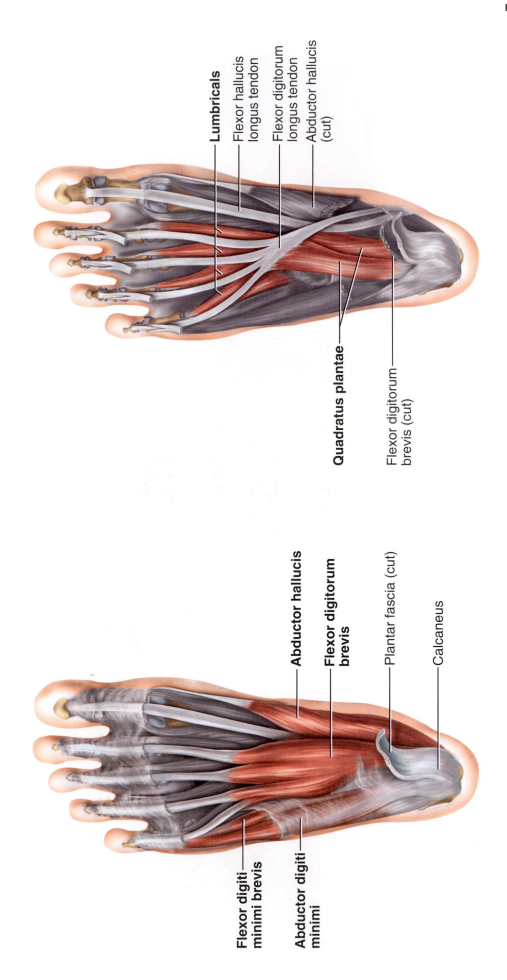

Lumbricals

Flexor hallucis longus tendon

Flexor digitorum longus tendon

Abductor hallucis (cut)

Quadratus plantae

Flexor digitorum brevis (cut)

Abductor hallucis

Flexor digitorum brevis

Plantar fascia (cut)

Calcaneus

Flexor digiti minimi brevis

Abductor digiti minimi

Figure 10.40a and b   Intrinsic Muscles of the Foot—Layers 1 and 2

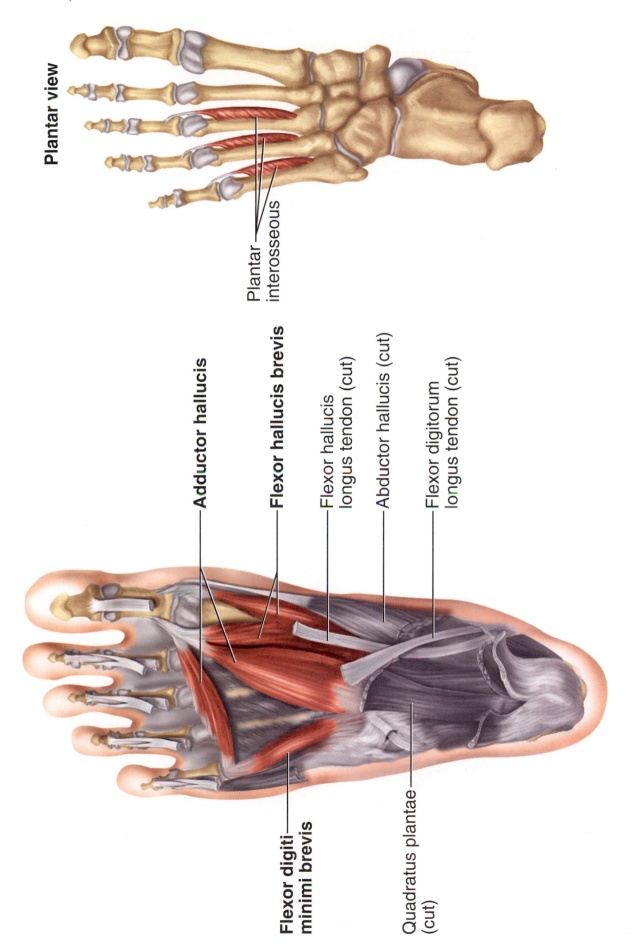

Plantar view

Plantar interosseous

Adductor hallucis

Flexor hallucis brevis

Flexor hallucis longus tendon (cut)

Abductor hallucis (cut)

Flexor digitorum longus tendon (cut)

Flexor digiti minimi brevis

Quadratus plantae (cut)

Figure 10.40c and d   Intrinsic Muscles of the Foot—Layers 3 and 4

Unit 5: The Spinal Cord and Spinal Nerves

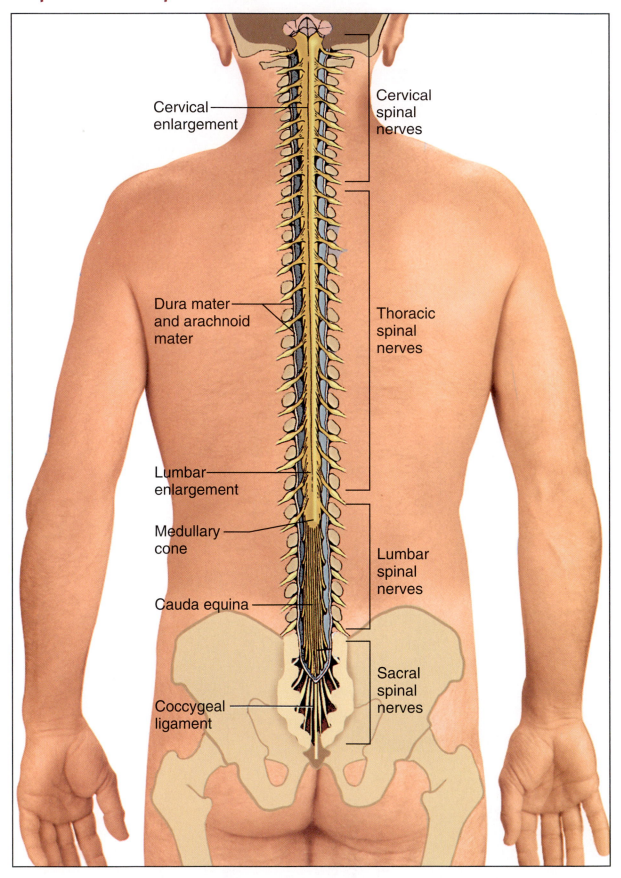

Cervical enlargement

Cervical spinal nerves

Dura mater and arachnoid mater

Thoracic spinal nerves

Lumbar enlargement

Medullary cone

Lumbar spinal nerves

Cauda equina

Coccygeal ligament

Sacral spinal nerves

Figure 13.1   The Spinal Cord, Dorsal Aspect

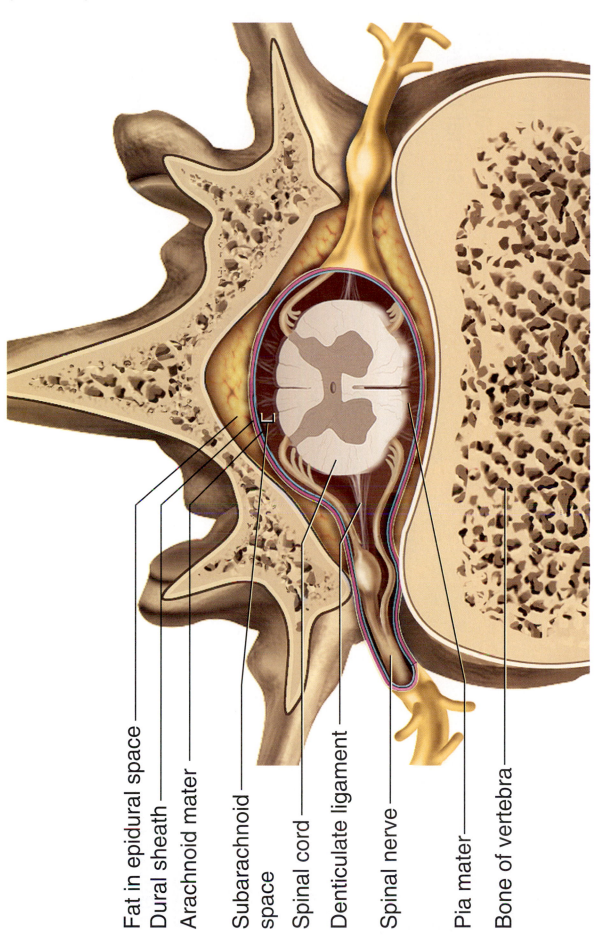

Fat in epidural space

Dural sheath

Arachnoid mater

Subarachnoid space

Spinal cord

Denticulate ligament

Spinal nerve

Pia mater

Bone of vertebra

Figure 13.2a Relationship to the Vertebra, Meninges, and Spinal Nerve

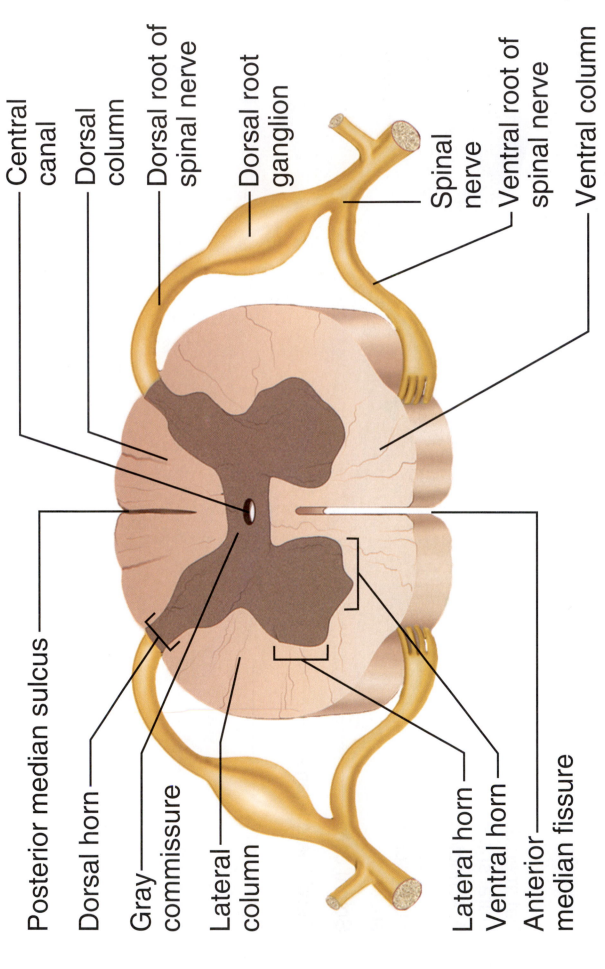

Central canal

Dorsal column

Dorsal root of spinal nerve

Dorsal root ganglion

Spinal nerve

Ventral root of spinal nerve

Ventral column

Posterior median sulcus

Dorsal horn

Gray commissure

Lateral column

Lateral horn

Ventral horn

Anterior median fissure

Figure 13.2b   Cross Section of the Thoracic Spinal Cord

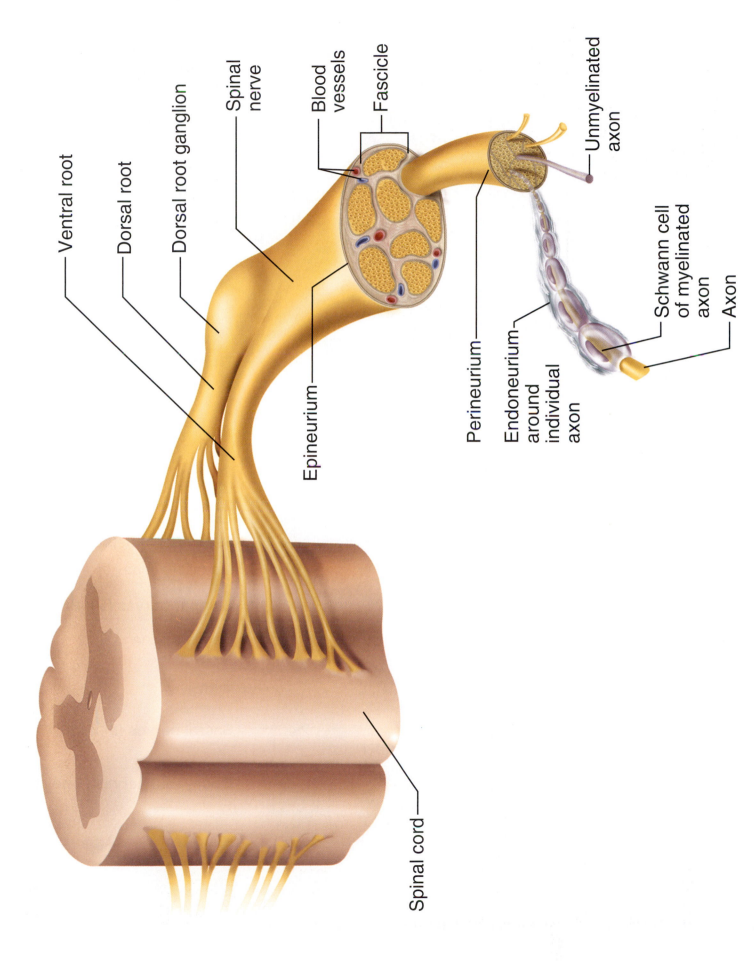

Ventral root

Dorsal root

Dorsal root ganglion

Spinal nerve

Blood vessels

Fascicle

Unmyelinated axon

Epineurium

Perineurium

Endoneurium around individual axon

Schwann cell of myelinated axon

Axon

Spinal cord

Figure 13.8a   Anatomy of a Nerve

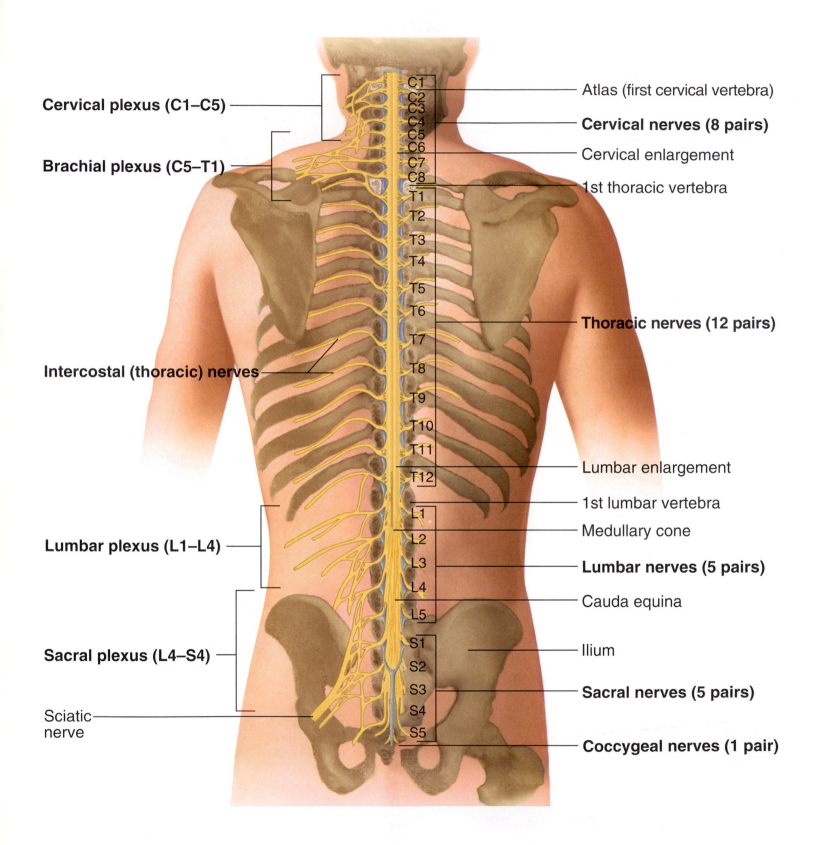

Cervical plexus (C1–C5)

Brachial plexus (C5–T1)

Intercostal (thoracic) nerves

Lumbar plexus (L1–L4)

Sacral plexus (L4–S4)

Sciatic
nerve

C1
C2
C3
C4
C5
C6
C7
C8
T1
T2
T3
T4
T5
T6
T7
T8
T9
T10
T11
T12
L1
L2
L3
L4
L5
S1
S2
S3
S4
S5

Atlas (first cervical vertebra)

Cervical nerves (8 pairs)

Cervical enlargement

1st thoracic vertebra

Thoracic nerves (12 pairs)

Lumbar enlargement

1st lumbar vertebra

Medullary cone

Lumbar nerves (5 pairs)

Cauda equina

Ilium

Sacral nerves (5 pairs)

Coccygeal nerves (1 pair)

Figure 13.10   The Spinal Nerve Roots and Plexuses, Dorsal View

Posterior

Anterior

Spine of vertebra

Deep muscles of back

Spinal cord

Spinal nerve

Meningeal branch

Communicating rami

Sympathetic ganglion

Dorsal root

Dorsal root ganglion

Dorsal ramus

Ventral ramus

Ventral root

Body of vertebra

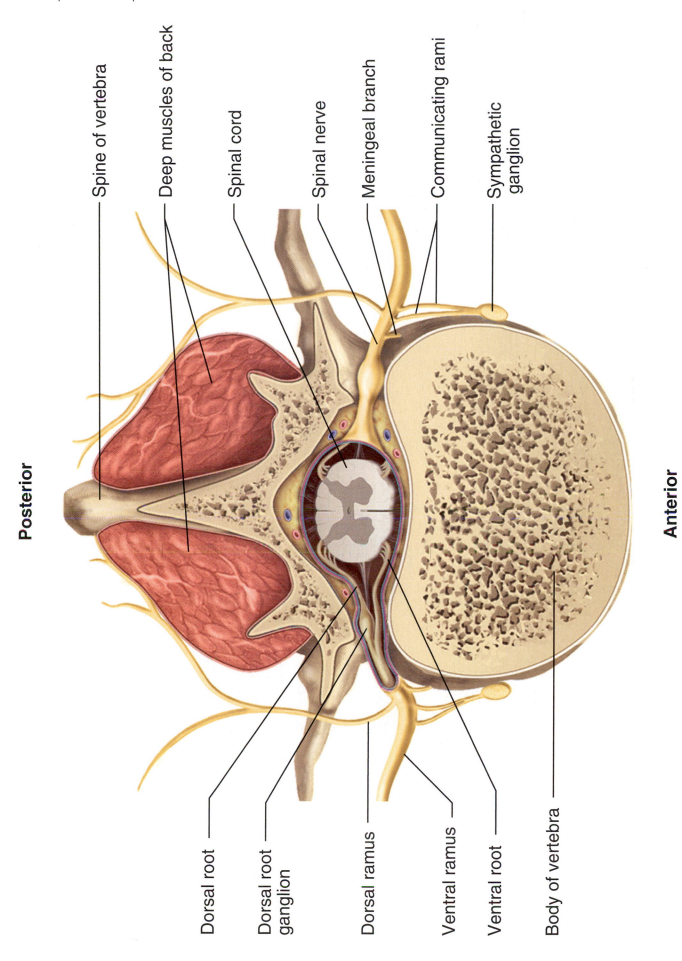

Figure 13.11   **Branches of the Spinal Nerve in Relation to the Spinal Cord and Vertebra**

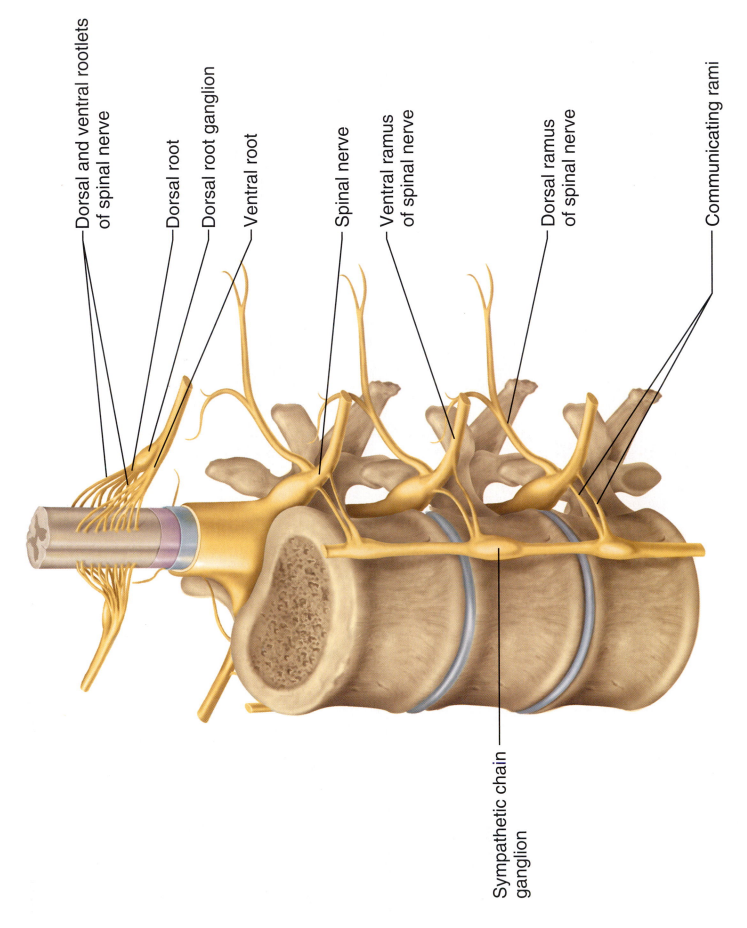

Dorsal and ventral rootlets of spinal nerve

Dorsal root

Dorsal root ganglion

Ventral root

Spinal nerve

Ventral ramus of spinal nerve

Dorsal ramus of spinal nerve

Communicating rami

Sympathetic chain ganglion

Figure 13.13a Anterolateral View of the Spinal Nerves

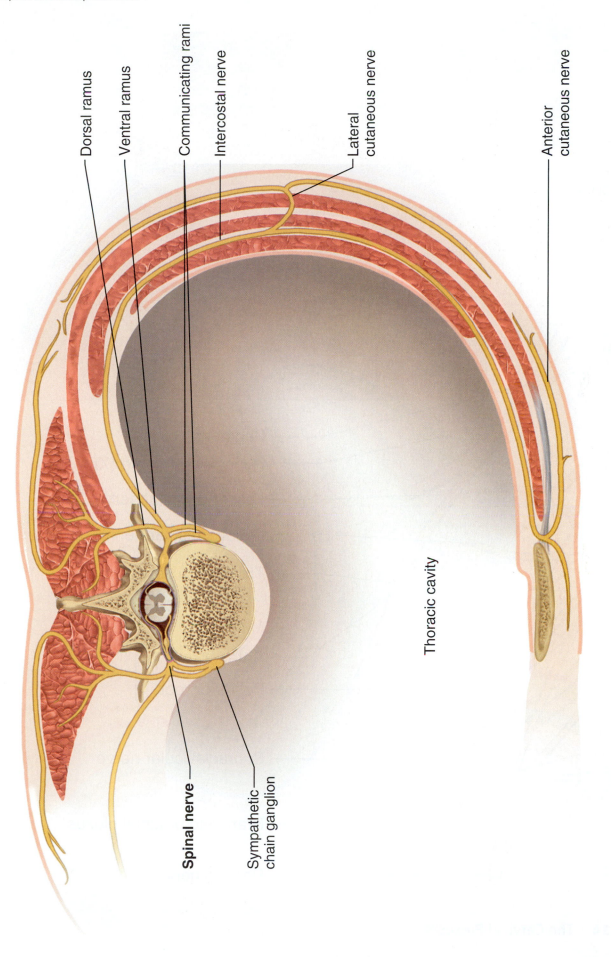

Dorsal ramus

Ventral ramus

Communicating rami

Intercostal nerve

Lateral cutaneous nerve

Anterior cutaneous nerve

Thoracic cavity

Spinal nerve

Sympathetic chain ganglion

Figure 13.13b Cross Section of the Thorax

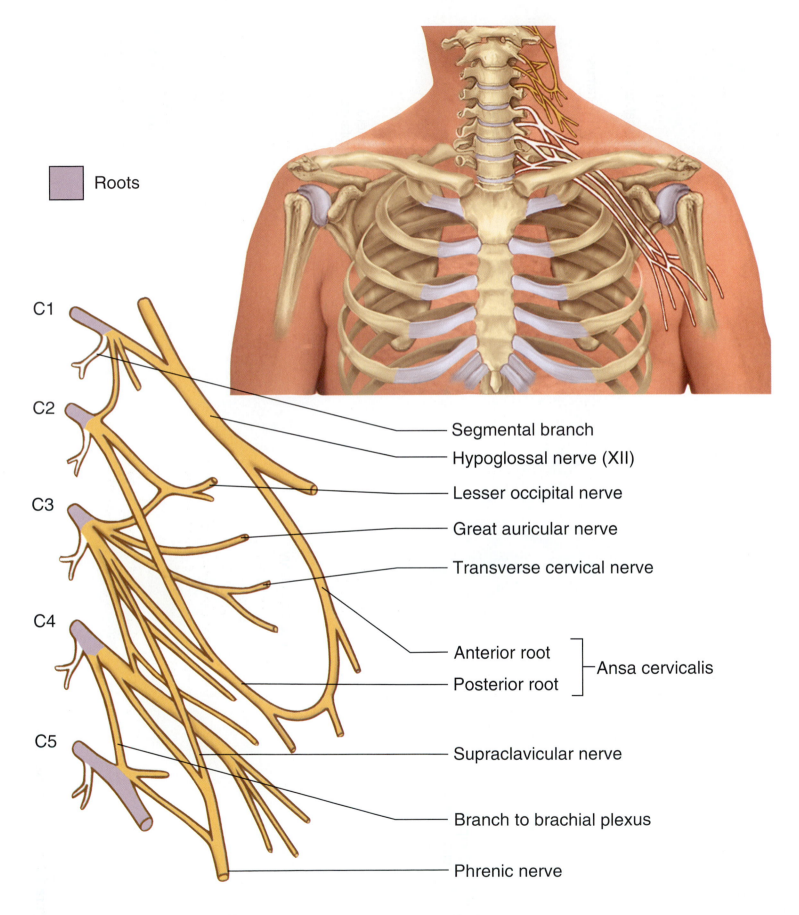

Roots

C1

C2

C3

C4

C5

Segmental branch

Hypoglossal nerve (XII)

Lesser occipital nerve

Great auricular nerve

Transverse cervical nerve

Anterior root

Posterior root

Ansa cervicalis

Supraclavicular nerve

Branch to brachial plexus

Phrenic nerve

Figure 13.14   The Cervical Plexus

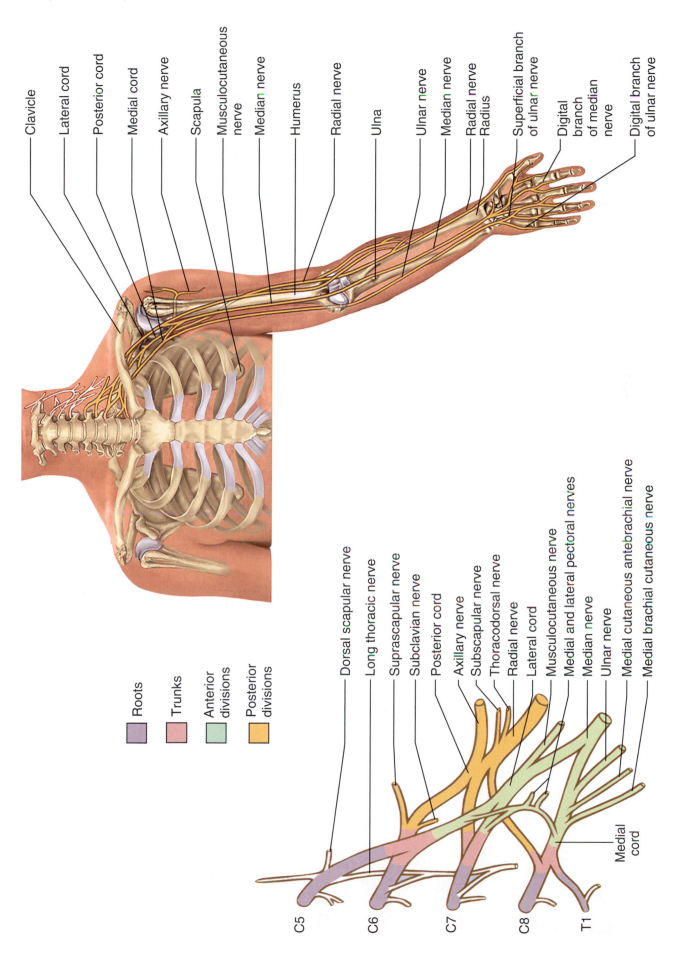

Clavicle

Lateral cord

Posterior cord

Medial cord

Axillary nerve

Scapula

Musculocutaneous nerve

Median nerve

Humerus

Radial nerve

Ulna

Ulnar nerve

Median nerve

Radial nerve

Radius

Superficial branch of ulnar nerve

Digital branch of median nerve

Digital branch of ulnar nerve

Roots

Trunks

Anterior divisions

Posterior divisions

Dorsal scapular nerve

Long thoracic nerve

Suprascapular nerve

Subclavian nerve

Posterior cord

Axillary nerve

Subscapular nerve

Thoracodorsal nerve

Radial nerve

Lateral cord

Musculocutaneous nerve

Medial and lateral pectoral nerves

Median nerve

Ulnar nerve

Medial cutaneous antebrachial nerve

Medial brachial cutaneous nerve

Medial cord

C5

C6

C7

C8

T1

Figure 13.15   The Brachial Plexus

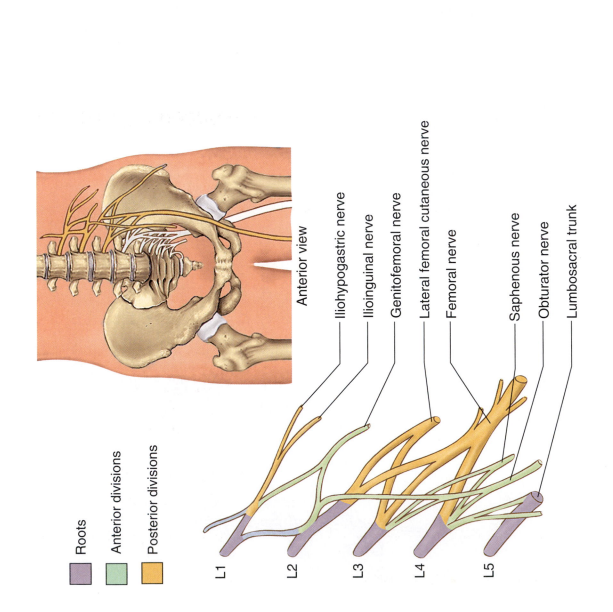

From lumbar plexus

From sacral plexus

Os coxae

Sacrum

Femoral nerve

Pudendal nerve

Sciatic nerve

Femur

Tibial nerve

Common fibular nerve

Superficial fibular nerve

Deep fibular nerve

Fibula

Tibia

Tibial nerve

Medial plantar nerve

Lateral plantar nerve

Posterior view

Anterior view

Iliohypogastric nerve

Ilioinguinal nerve

Genitofemoral nerve

Lateral femoral cutaneous nerve

Femoral nerve

Saphenous nerve

Obturator nerve

Lumbosacral trunk

Roots

Anterior divisions

Posterior divisions

L1

L2

L3

L4

L5

Figure 13.17   **The Lumbar Plexus**

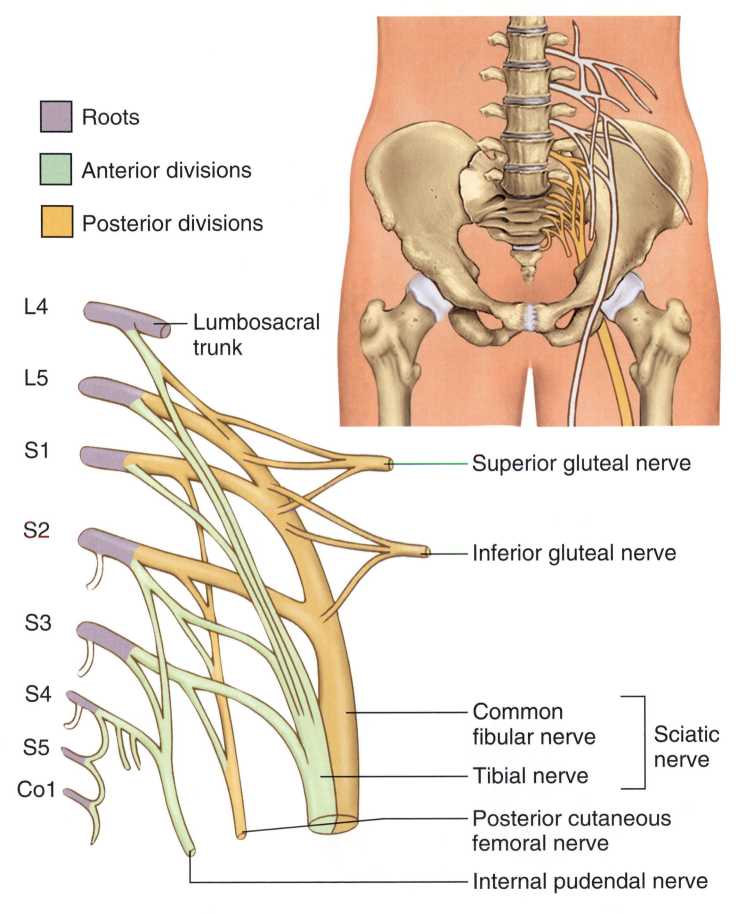

Roots

Anterior divisions

Posterior divisions

L4

Lumbosacral trunk

L5

S1 — Superior gluteal nerve

S2 — Inferior gluteal nerve

S3

S4

S5

Co1

Common fibular nerve — Sciatic nerve

Tibial nerve

Posterior cutaneous femoral nerve

Internal pudendal nerve

Figure 13.18   The Sacral and Coccygeal Plexus

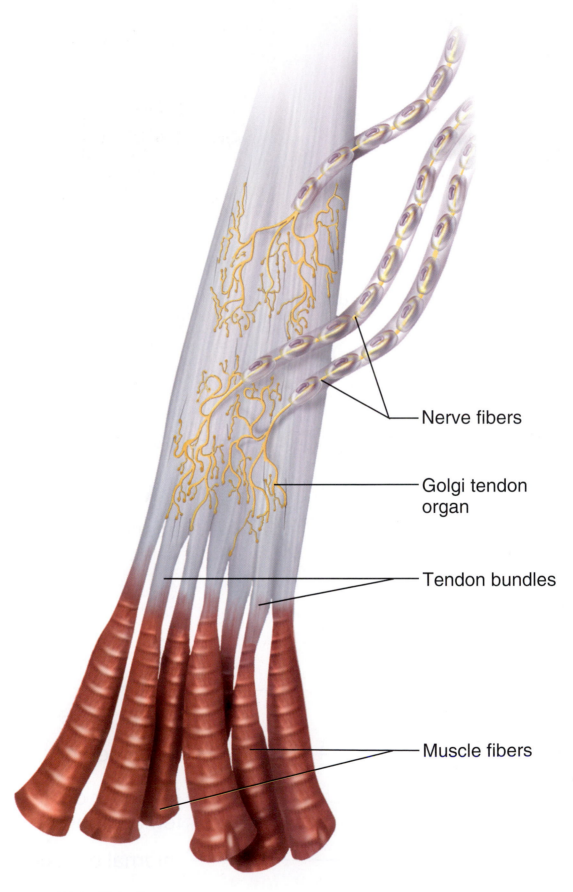

Nerve fibers

Golgi tendon
organ

Tendon bundles

Muscle fibers

Figure 13.23   A Golgi Tendon Organ

Unit 6: The Brain and Cranial Nerves

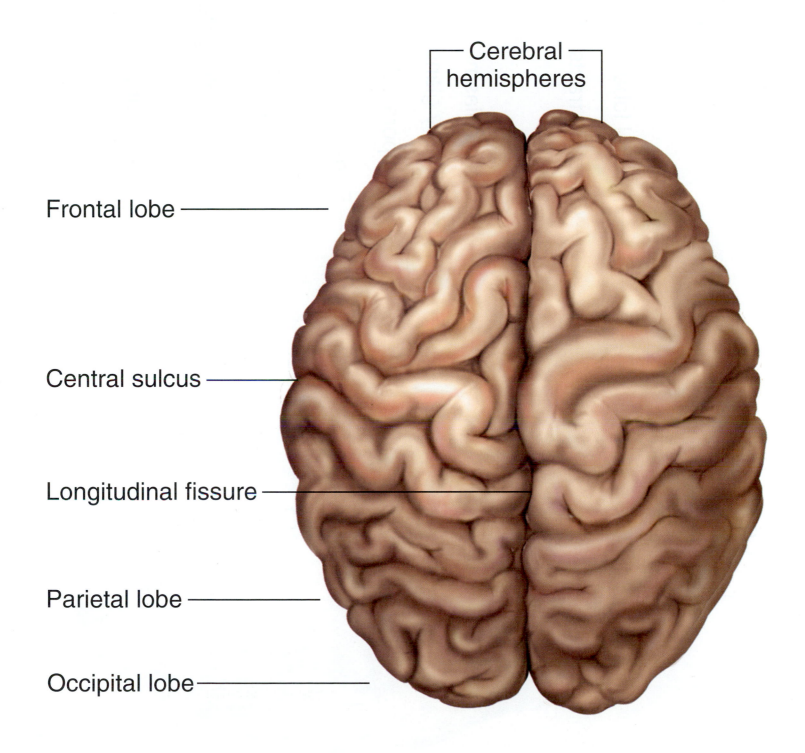

Cerebral hemispheres

Frontal lobe

Central sulcus

Longitudinal fissure

Parietal lobe

Occipital lobe

Figure 14.1a   Four Views of the Brain—Superior View

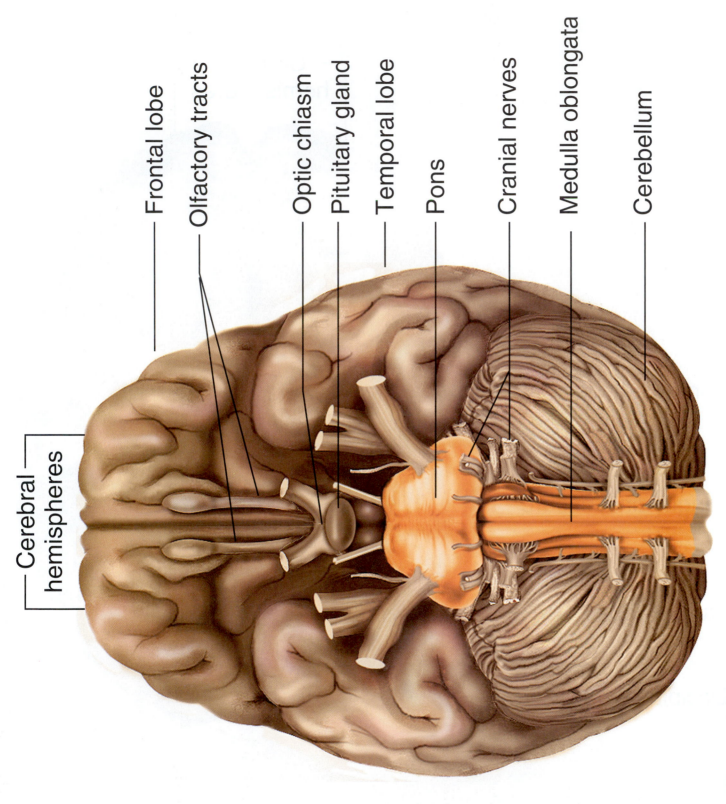

Frontal lobe

Olfactory tracts

Optic chiasm

Pituitary gland

Temporal lobe

Pons

Cranial nerves

Medulla oblongata

Cerebellum

Cerebral hemispheres

Figure 14.1b   Four Views of the Brain—Inferior View

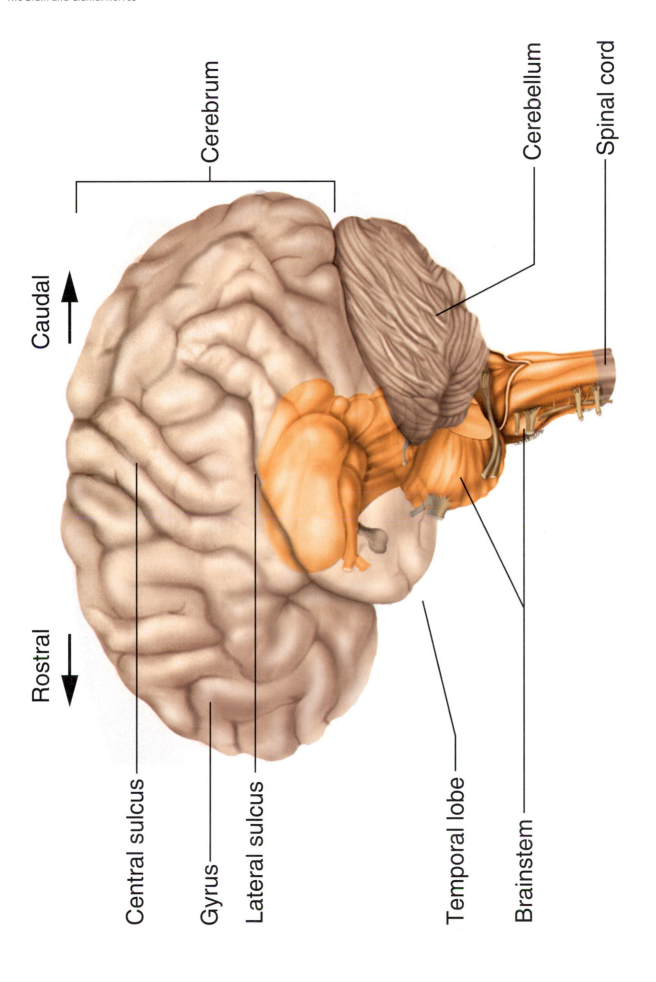

Cerebrum

Cerebellum

Spinal cord

Caudal

Rostral

Central sulcus

Gyrus

Lateral sulcus

Temporal lobe

Brainstem

Figure 14.1c Four Views of the Brain—Lateral View

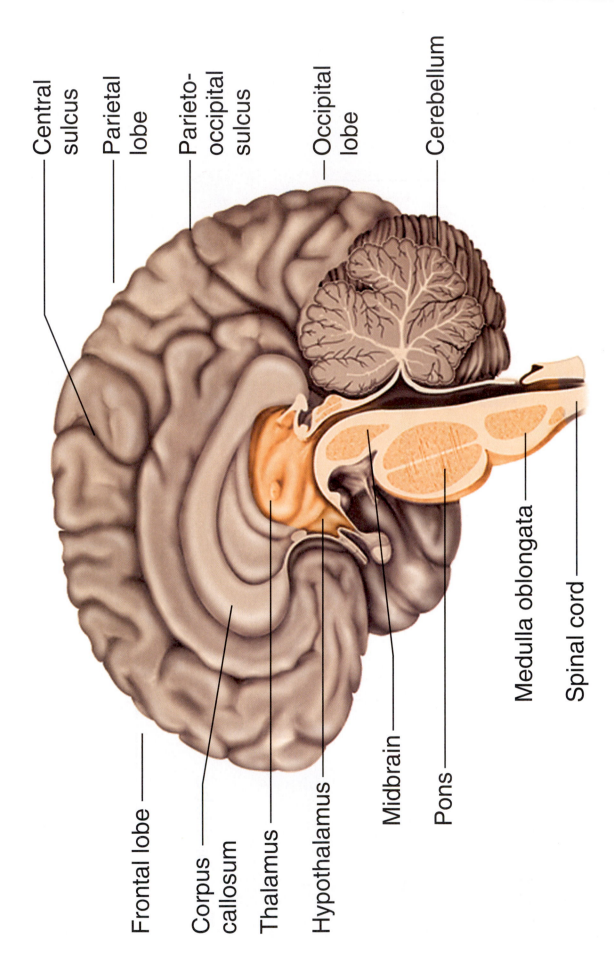

Central sulcus

Parietal lobe

Parieto-occipital sulcus

Occipital lobe

Cerebellum

Frontal lobe

Corpus callosum

Thalamus

Hypothalamus

Midbrain

Pons

Medulla oblongata

Spinal cord

Figure 14.1d   Four Views of the Brain—Median View

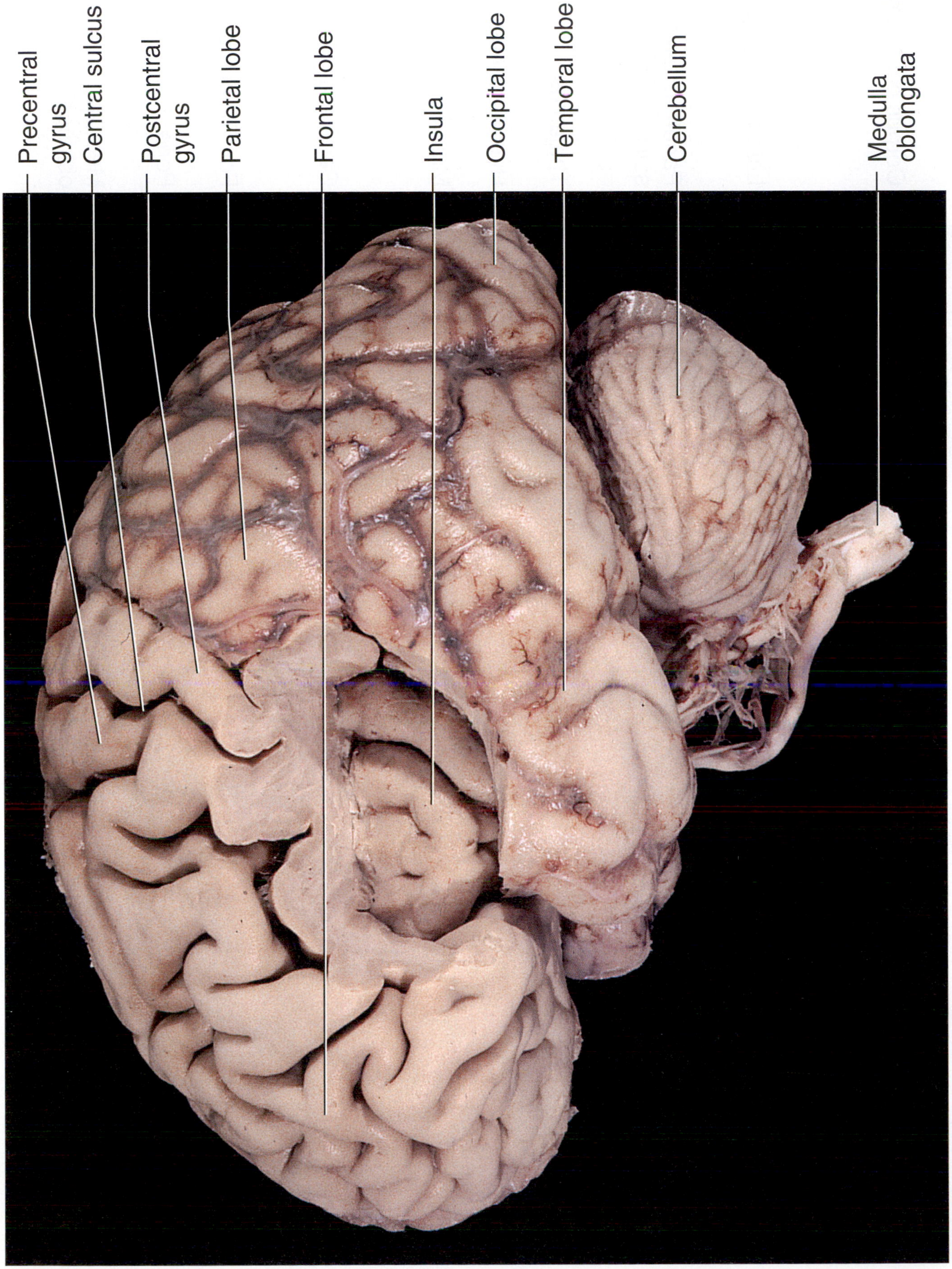

Precentral gyrus

Central sulcus

Postcentral gyrus

Parietal lobe

Frontal lobe

Insula

Occipital lobe

Temporal lobe

Cerebellum

Medulla oblongata

Figure 14.2a   Dissections of the Brain—Left Lateral View

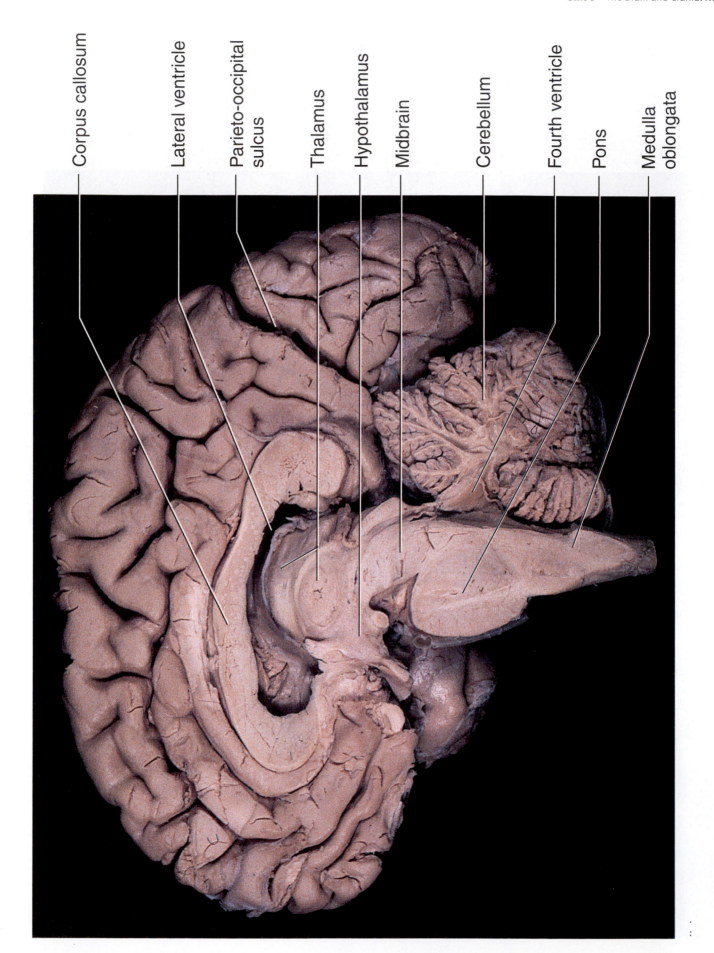

Corpus callosum

Lateral ventricle

Parieto-occipital sulcus

Thalamus

Hypothalamus

Midbrain

Cerebellum

Fourth ventricle

Pons

Medulla oblongata

Figure 14.2b   Dissections of the Brain—Median Section, Left Lateral View

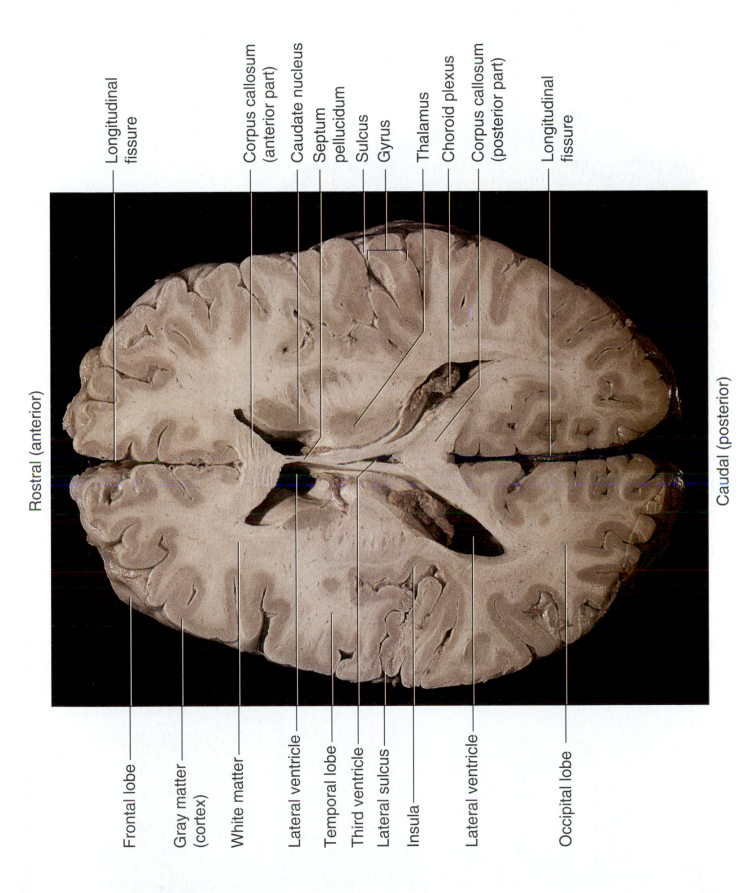

Figure 14.6c   Ventricles of the Brain

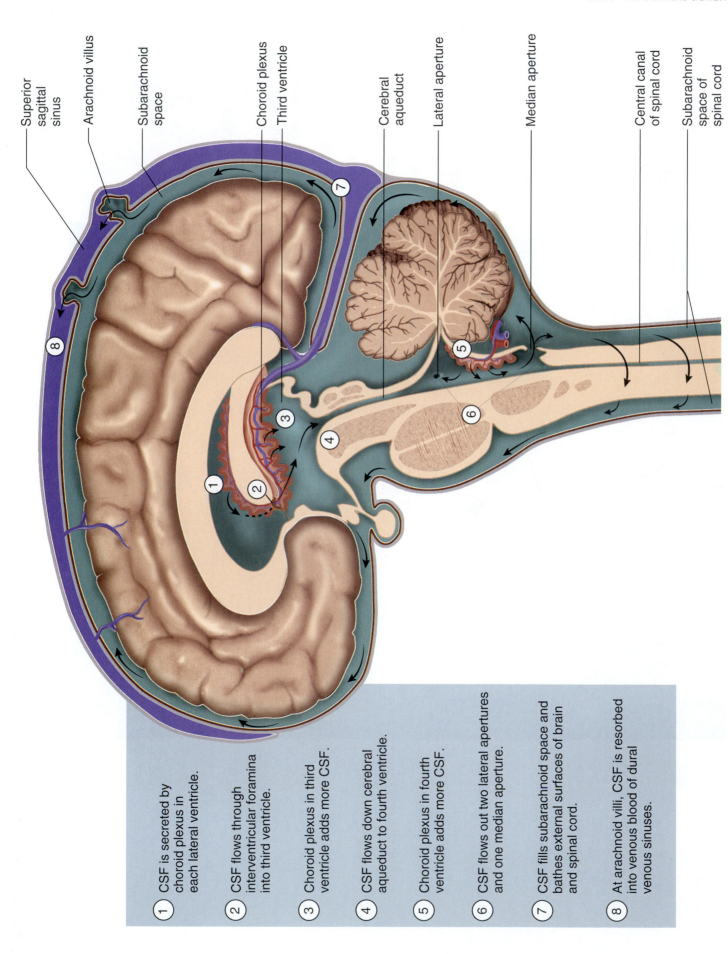

Superior sagittal sinus

Arachnoid villus

Subarachnoid space

Choroid plexus

Third ventricle

Cerebral aqueduct

Lateral aperture

Median aperture

Central canal of spinal cord

Subarachnoid space of spinal cord

① CSF is secreted by choroid plexus in each lateral ventricle.

② CSF flows through interventricular foramina into third ventricle.

③ Choroid plexus in third ventricle adds more CSF.

④ CSF flows down cerebral aqueduct to fourth ventricle.

⑤ Choroid plexus in fourth ventricle adds more CSF.

⑥ CSF flows out two lateral apertures and one median aperture.

⑦ CSF fills subarachnoid space and bathes external surfaces of brain and spinal cord.

⑧ At arachnoid villi, CSF is resorbed into venous blood of dural venous sinuses.

Figure 14.7   Flow of Cerebrospinal Fluid

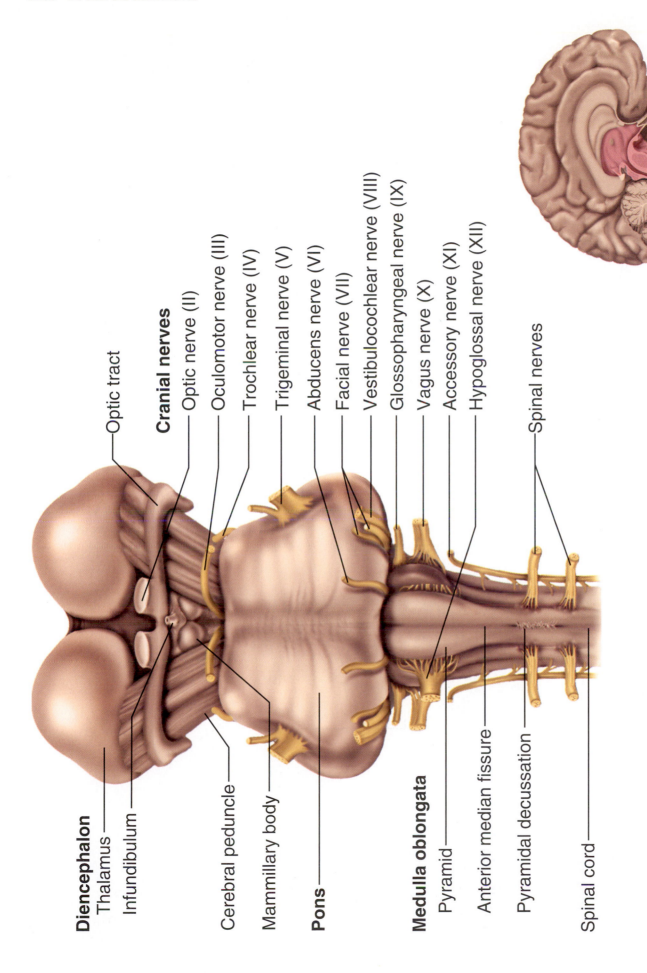

Diencephalon
Thalamus
Infundibulum

Optic tract

Cranial nerves

Optic nerve (II)
Oculomotor nerve (III)
Trochlear nerve (IV)
Trigeminal nerve (V)
Abducens nerve (VI)
Facial nerve (VII)
Vestibulocochlear nerve (VIII)
Glossopharyngeal nerve (IX)
Vagus nerve (X)
Accessory nerve (XI)
Hypoglossal nerve (XII)

Spinal nerves

Cerebral peduncle

Mammillary body

Pons

Medulla oblongata
Pyramid
Anterior median fissure
Pyramidal decussation

Spinal cord

Figure 14.8a The Brainstem—Ventral Aspect

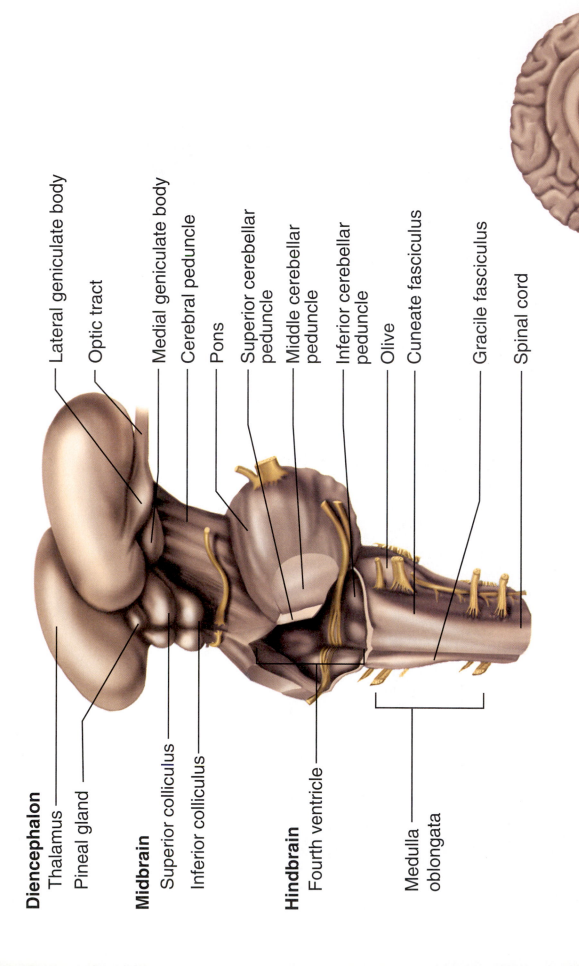

Diencephalon
Thalamus
Pineal gland

Midbrain
Superior colliculus
Inferior colliculus

Hindbrain
Fourth ventricle

Medulla
oblongata

Lateral geniculate body
Optic tract
Medial geniculate body
Cerebral peduncle
Pons
Superior cerebellar peduncle
Middle cerebellar peduncle
Inferior cerebellar peduncle
Olive
Cuneate fasciculus
Gracile fasciculus
Spinal cord

Figure 14.8b   The Brainstem—Right Dorsolateral Aspect

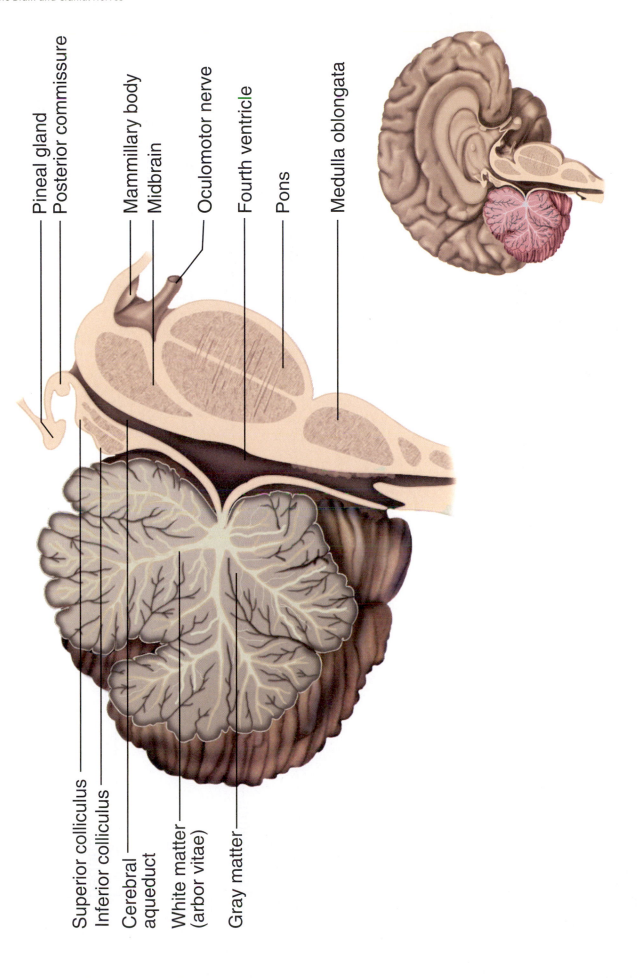

Pineal gland
Posterior commissure
Mammillary body
Midbrain
Oculomotor nerve
Fourth ventricle
Pons
Medulla oblongata

Superior colliculus
Inferior colliculus
Cerebral aqueduct
White matter (arbor vitae)
Gray matter

Figure 14.9a The Cerebellum

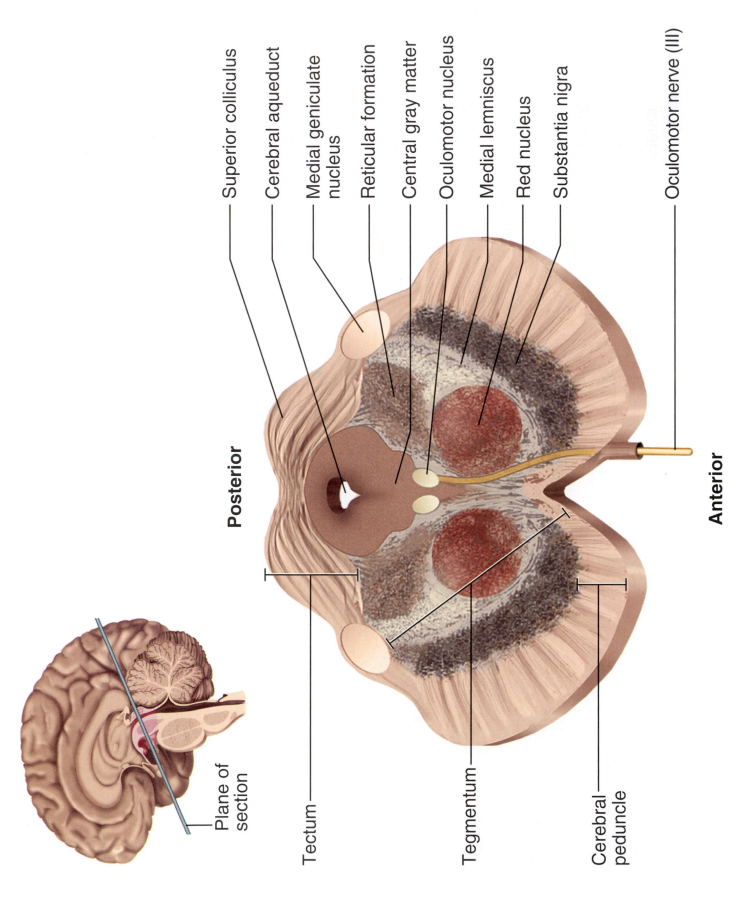

Superior colliculus

Cerebral aqueduct

Medial geniculate nucleus

Reticular formation

Central gray matter

Oculomotor nucleus

Medial lemniscus

Red nucleus

Substantia nigra

Oculomotor nerve (III)

Posterior

Anterior

Plane of section

Tectum

Tegmentum

Cerebral peduncle

Figure 14.10   Cross Section of the Midbrain

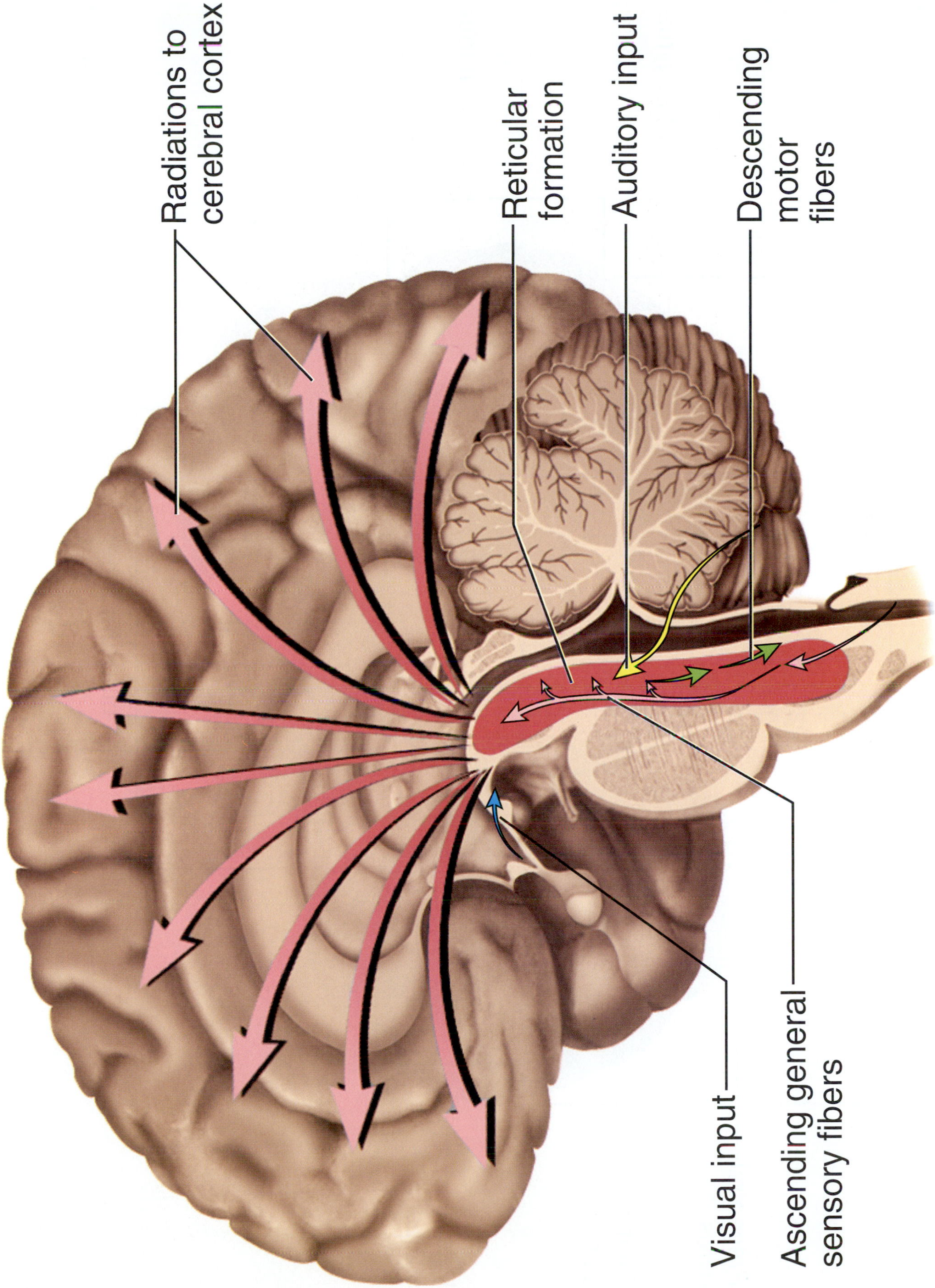

Radiations to cerebral cortex

Reticular formation

Auditory input

Descending motor fibers

Visual input

Ascending general sensory fibers

Figure 14.11   The Reticular Formation

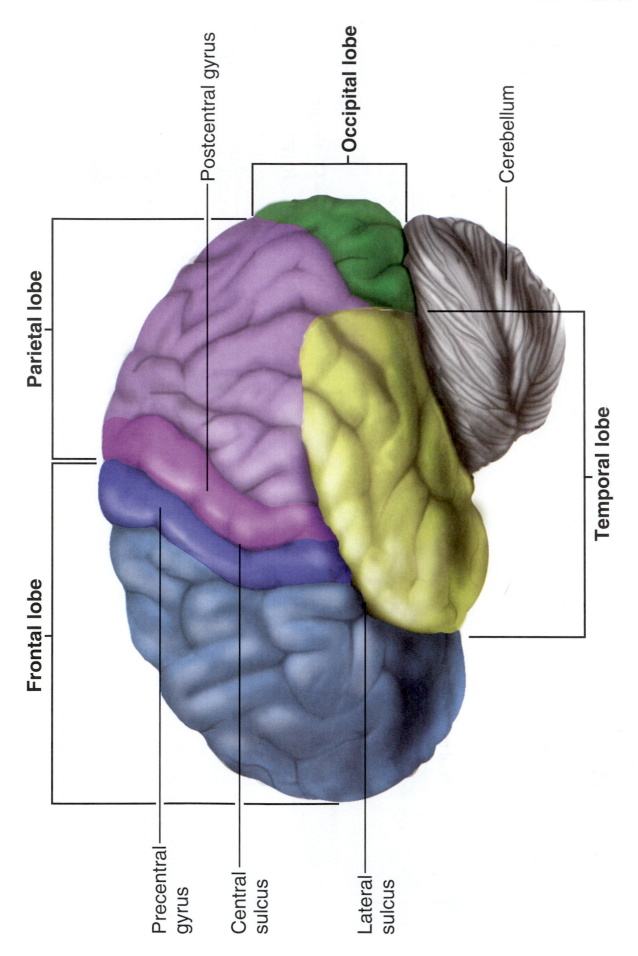

Postcentral gyrus

Occipital lobe

Cerebellum

Parietal lobe

Temporal lobe

Frontal lobe

Precentral gyrus

Central sulcus

Lateral sulcus

Figure 14.13   Lobes of the Cerebrum

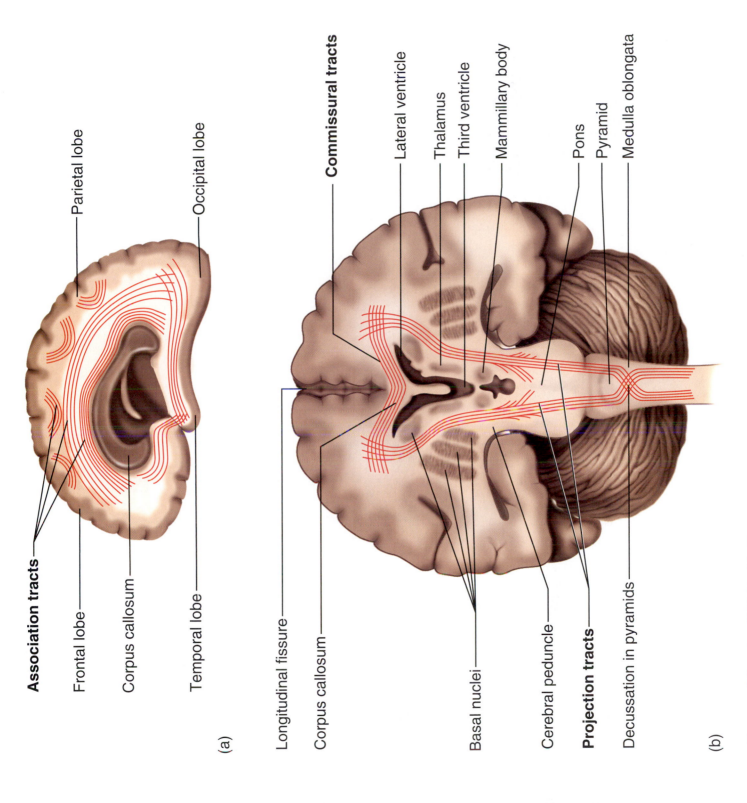

Association tracts

Parietal lobe

Occipital lobe

Frontal lobe

Corpus callosum

Temporal lobe

(a)

Commissural tracts

Lateral ventricle

Thalamus

Third ventricle

Mammillary body

Pons

Pyramid

Medulla oblongata

Longitudinal fissure

Corpus callosum

Basal nuclei

Cerebral peduncle

Projection tracts

Decussation in pyramids

(b)

Figure 14.14 Tracts of Cerebral White Matter

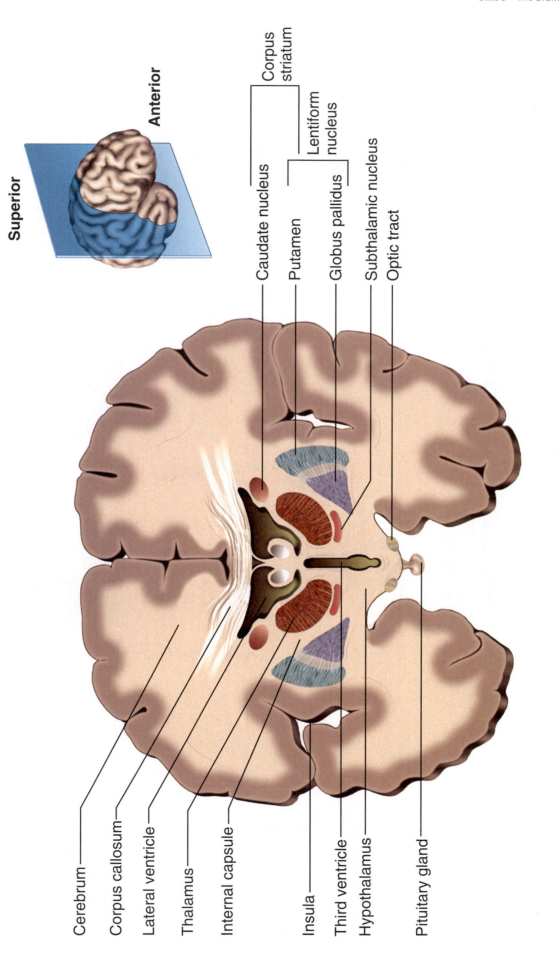

Superior

Anterior

Corpus
striatum

Caudate nucleus

Putamen

Lentiform
nucleus

Globus pallidus

Subthalamic nucleus

Optic tract

Cerebrum

Corpus callosum

Lateral ventricle

Thalamus

Internal capsule

Insula

Third ventricle

Hypothalamus

Pituitary gland

Figure 14.16   The Basal Nuclei

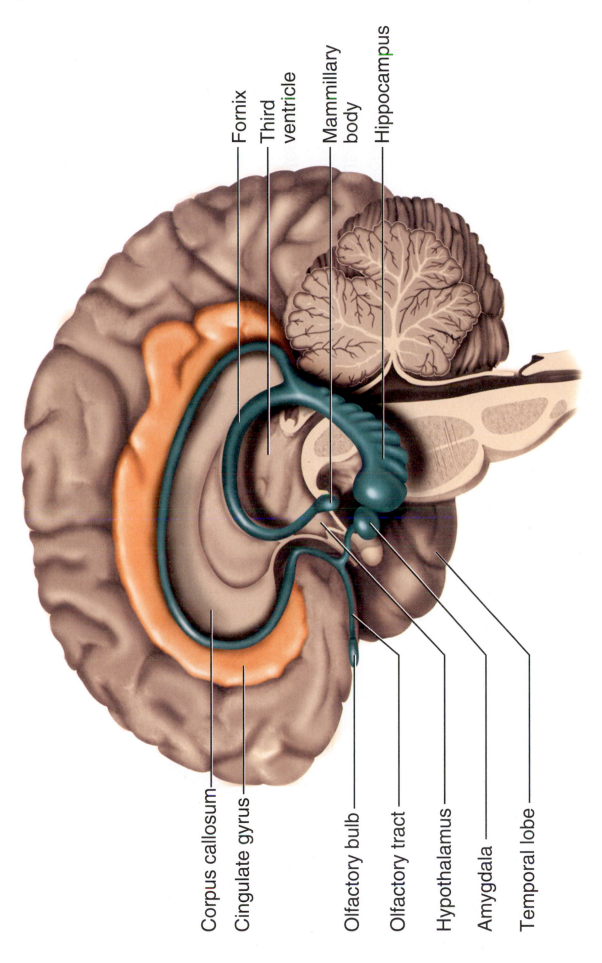

Fornix

Third ventricle

Mammillary body

Hippocampus

Corpus callosum

Cingulate gyrus

Olfactory bulb

Olfactory tract

Hypothalamus

Amygdala

Temporal lobe

Figure 14.17 The Limbic System

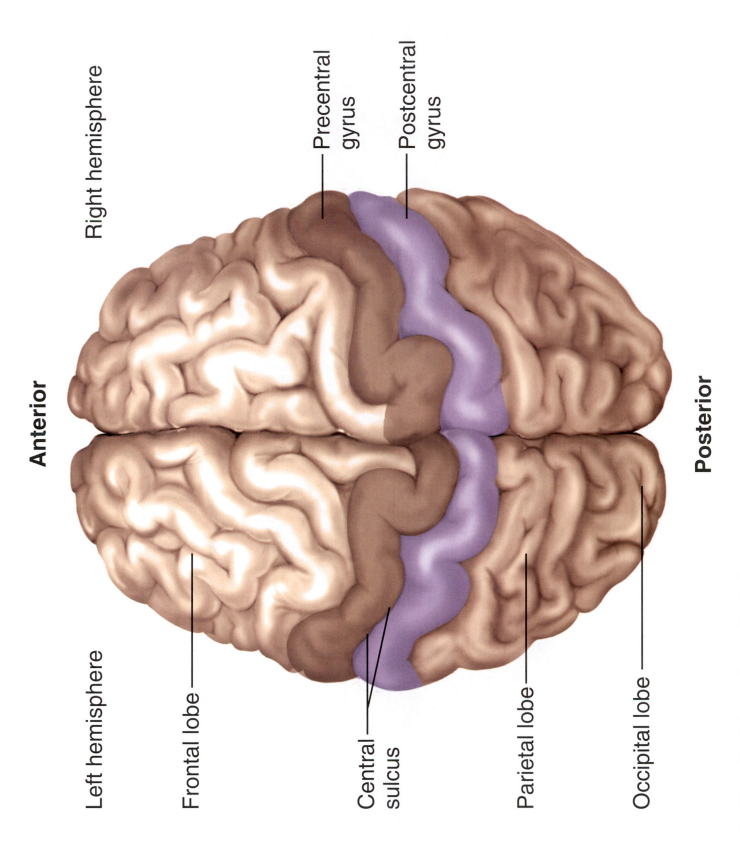

Right hemisphere

Precentral gyrus

Postcentral gyrus

Anterior

Posterior

Left hemisphere

Frontal lobe

Central sulcus

Parietal lobe

Occipital lobe

Figure 14.21a   The Primary Somesthetic Cortex

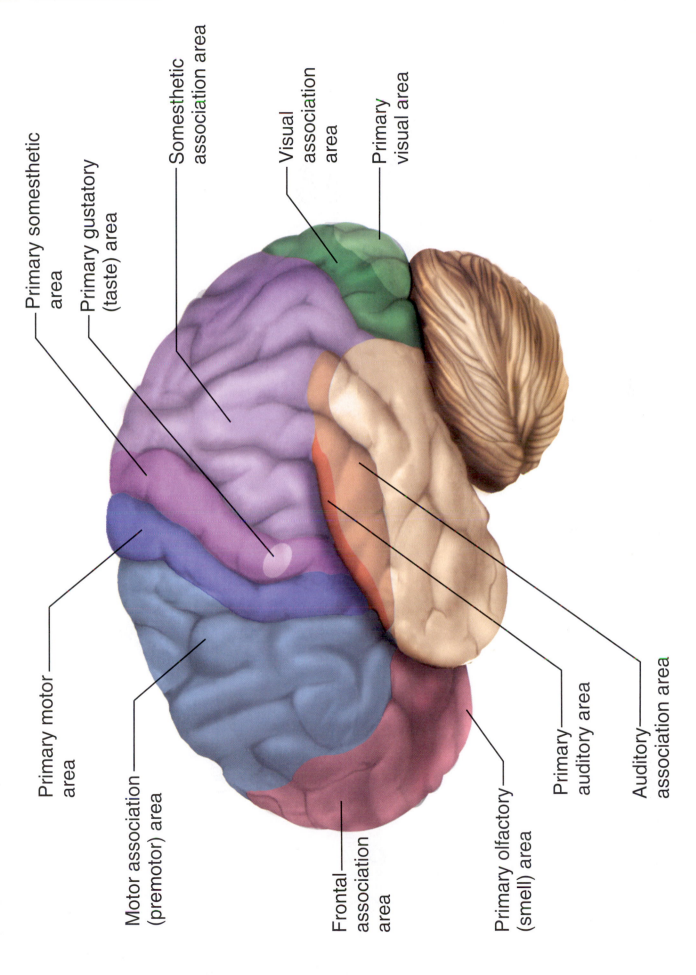

Primary somesthetic area

Primary gustatory (taste) area

Somesthetic association area

Visual association area

Primary visual area

Primary motor area

Motor association (premotor) area

Frontal association area

Primary olfactory (smell) area

Primary auditory area

Auditory association area

Figure 14.22   Some Functional Regions of the Cerebral Cortex

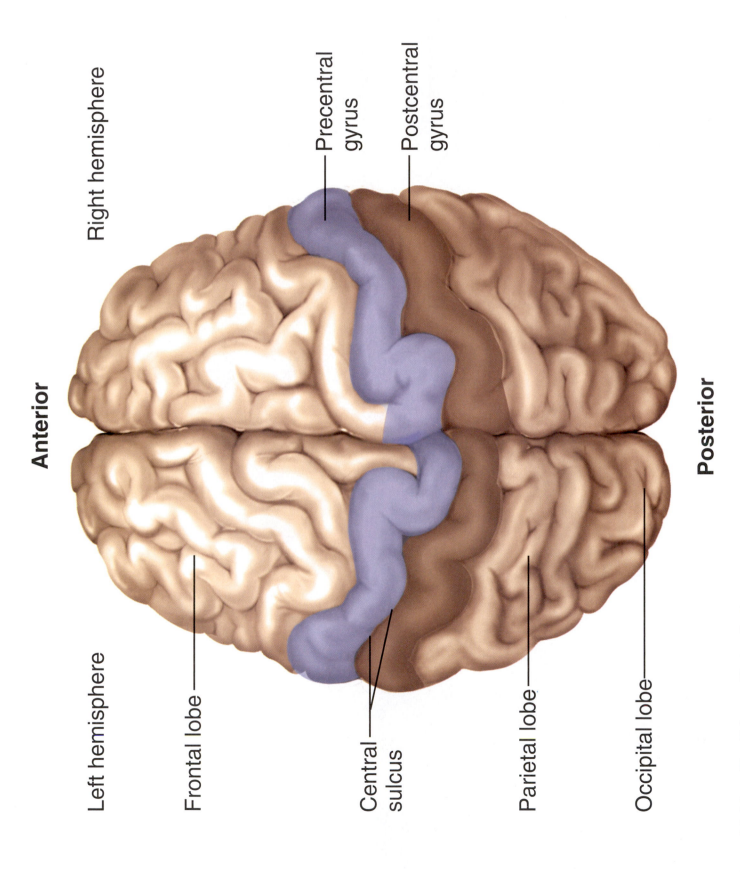

Figure 14.23a   The Primary Motor Cortex

Posterior

Anterior

Postcentral gyrus

Angular gyrus

Primary visual cortex

Wernicke's area

Precentral gyrus

Speech center of primary motor cortex

Primary auditory cortex (in lateral sulcus)

Broca's area

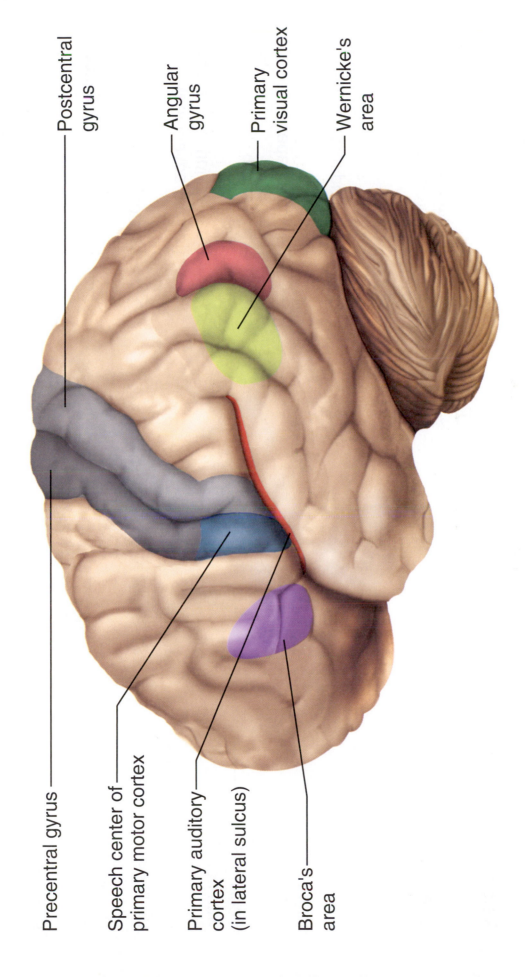

Figure 14.25 Language Centers of the Left Hemisphere

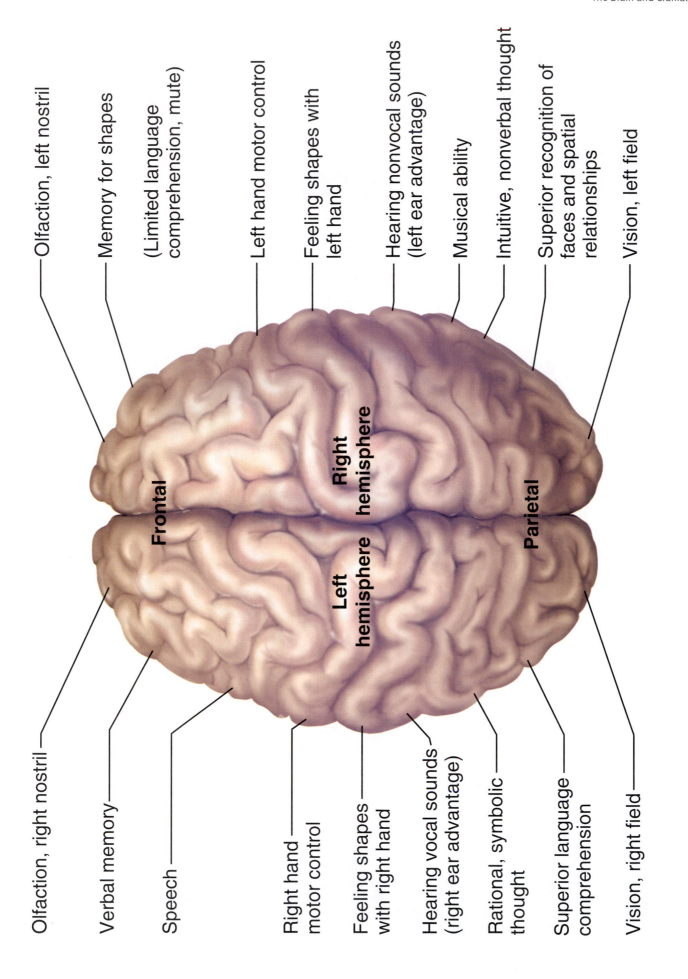

Olfaction, left nostril

Memory for shapes

(Limited language comprehension, mute)

Left hand motor control

Feeling shapes with left hand

Hearing nonvocal sounds (left ear advantage)

Musical ability

Intuitive, nonverbal thought

Superior recognition of faces and spatial relationships

Vision, left field

Frontal

Right hemisphere

Left hemisphere

Parietal

Olfaction, right nostril

Verbal memory

Speech

Right hand motor control

Feeling shapes with right hand

Hearing vocal sounds (right ear advantage)

Rational, symbolic thought

Superior language comprehension

Vision, right field

Figure 14.26   Lateralization of Cerebral Functions

Cranial nerves

Fibers of olfactory nerve (I)

Optic nerve (II)

Oculomotor nerve (III)

Trochlear nerve (IV)

Trigeminal nerve (V)

Abducens nerve (VI)

Facial nerve (VII)

Vestibulocochlear nerve (VIII)

Glossopharyngeal nerve (IX)

Vagus nerve (X)

Accessory nerve (XI)

Hypoglossal nerve (XII)

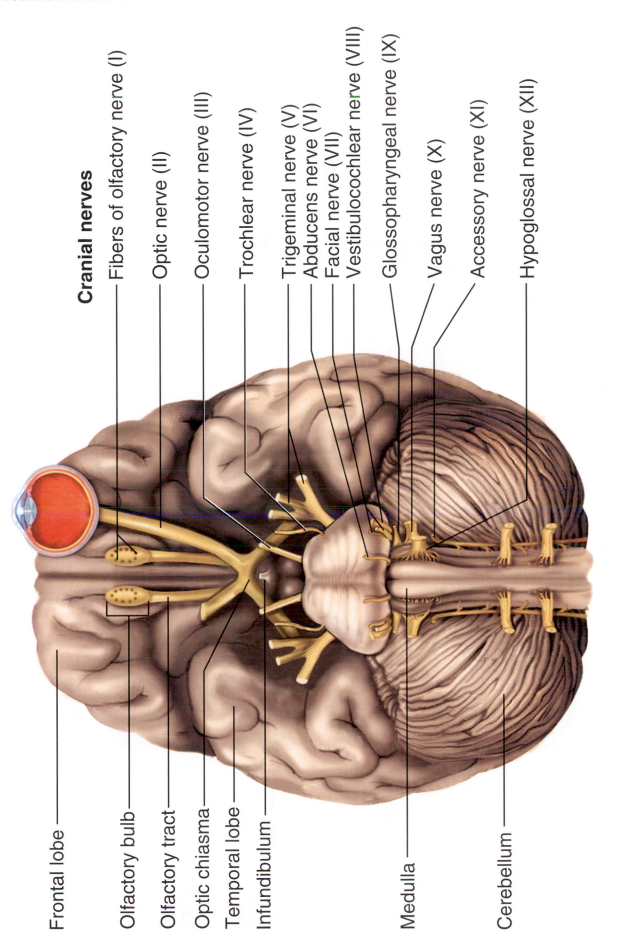

Frontal lobe

Olfactory bulb

Olfactory tract

Optic chiasma

Temporal lobe

Infundibulum

Medulla

Cerebellum

Figure 14.27a   The Cranial Nerves

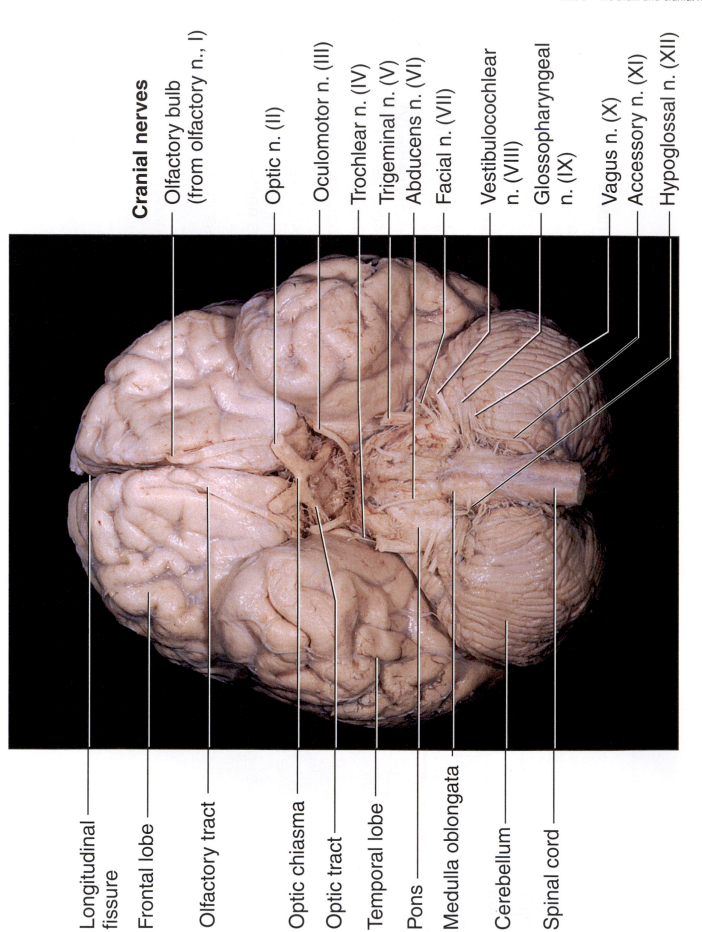

Cranial nerves

Olfactory bulb
(from olfactory n., I)

Optic n. (II)

Oculomotor n. (III)

Trochlear n. (IV)

Trigeminal n. (V)

Abducens n. (VI)

Facial n. (VII)

Vestibulocochlear
n. (VIII)

Glossopharyngeal
n. (IX)

Vagus n. (X)

Accessory n. (XI)

Hypoglossal n. (XII)

Longitudinal
fissure

Frontal lobe

Olfactory tract

Optic chiasma

Optic tract

Temporal lobe

Pons

Medulla oblongata

Cerebellum

Spinal cord

Figure 14.27b   The Cranial Nerves

Unit 7: The Sense Organs

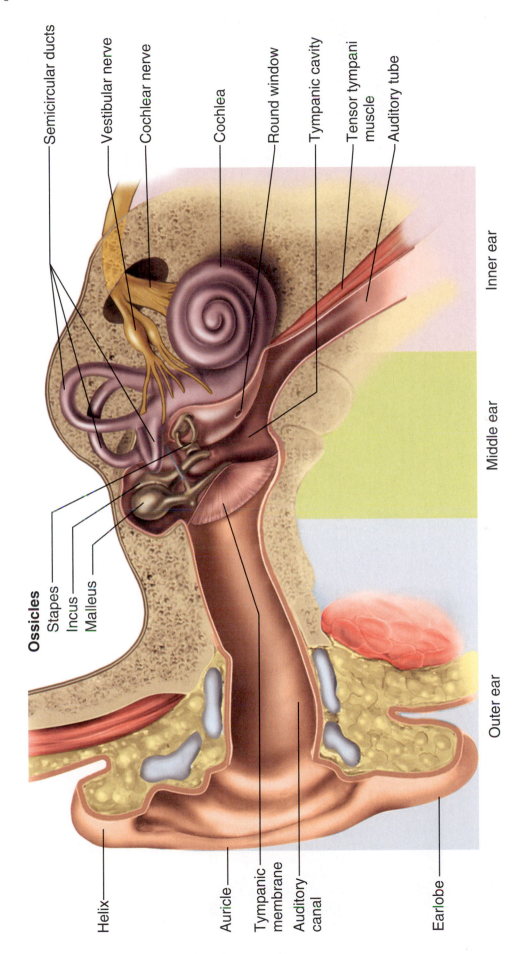

Semicircular ducts

Vestibular nerve

Cochlear nerve

Cochlea

Round window

Tympanic cavity

Tensor tympani muscle

Auditory tube

Inner ear

Middle ear

Outer ear

Ossicles
Stapes
Incus
Malleus

Helix

Auricle

Tympanic membrane

Auditory canal

Earlobe

Figure 16.10 Internal Anatomy of the Ear

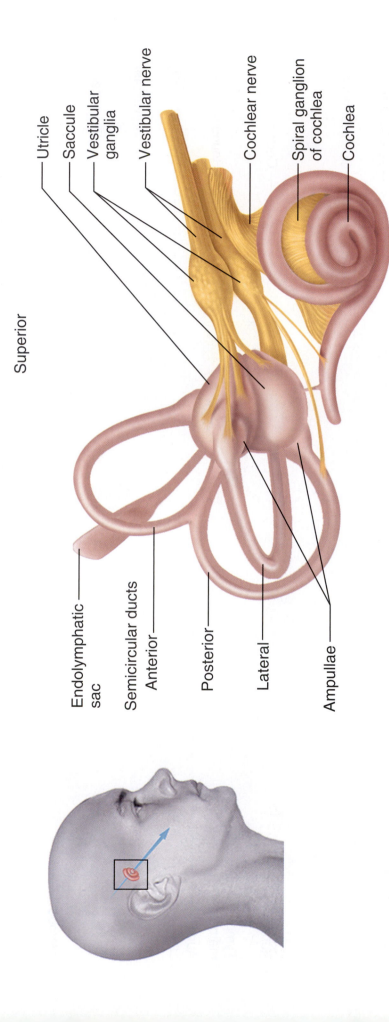

Superior

Utricle
Saccule
Vestibular ganglia
Vestibular nerve
Cochlear nerve
Spiral ganglion of cochlea
Cochlea

Endolymphatic sac
Semicircular ducts
Anterior
Posterior
Lateral
Ampullae

Figure 16.11a   Anatomy of the Inner Ear

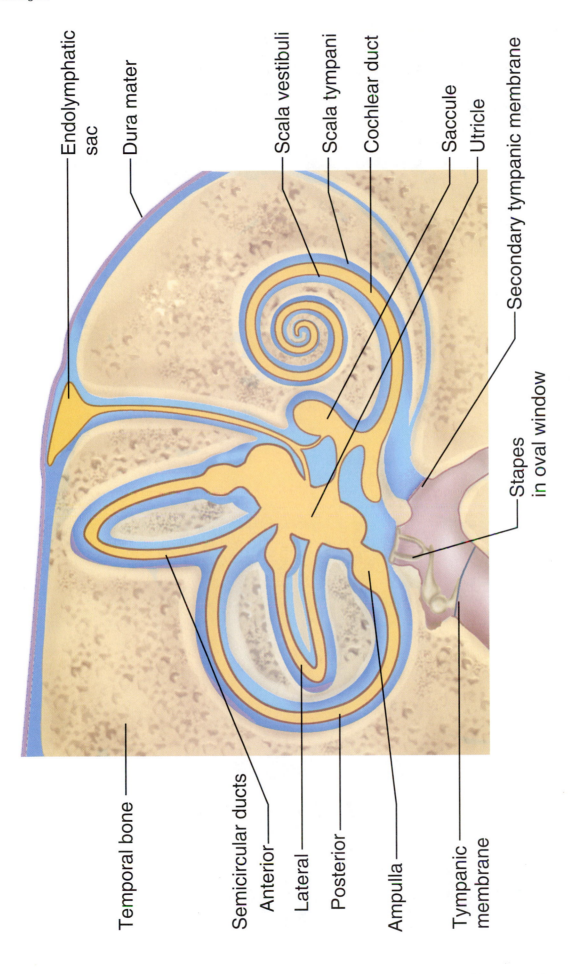

Endolymphatic sac

Dura mater

Scala vestibuli

Scala tympani

Cochlear duct

Saccule

Utricle

Secondary tympanic membrane

Stapes in oval window

Temporal bone

Semicircular ducts

Anterior

Lateral

Posterior

Ampulla

Tympanic membrane

Figure 16.11b   Anatomy of the Inner Ear

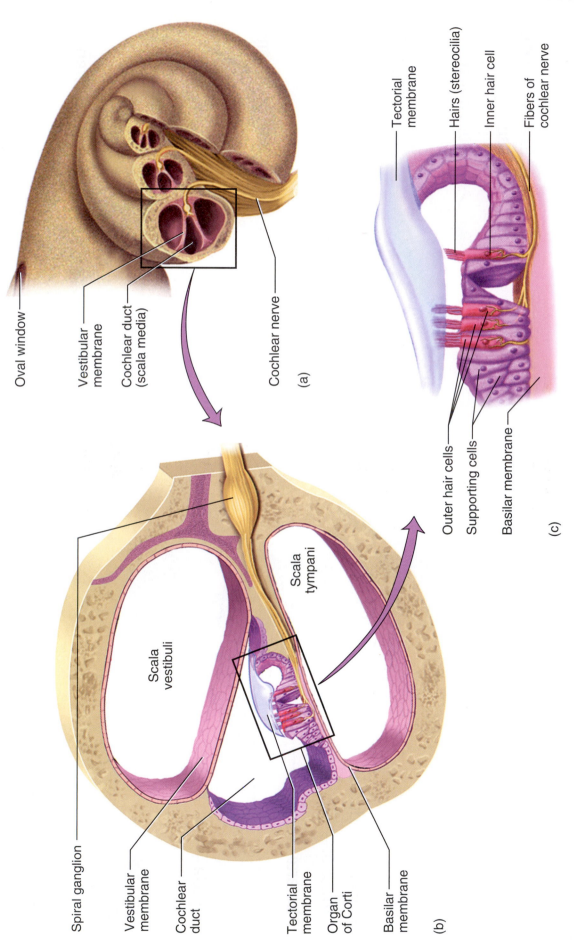

Oval window

Vestibular
membrane

Cochlear duct
(scala media)

Cochlear nerve

(a)

Tectorial
membrane

Hairs (stereocilia)

Inner hair cell

Fibers of
cochlear nerve

Outer hair cells

Supporting cells

Basilar membrane

(c)

Scala
vestibuli

Scala
tympani

Spiral ganglion

Vestibular
membrane

Cochlear
duct

Tectorial
membrane

Organ
of Corti

Basilar
membrane

(b)

Figure 16.12   **Anatomy of the Cochlea**

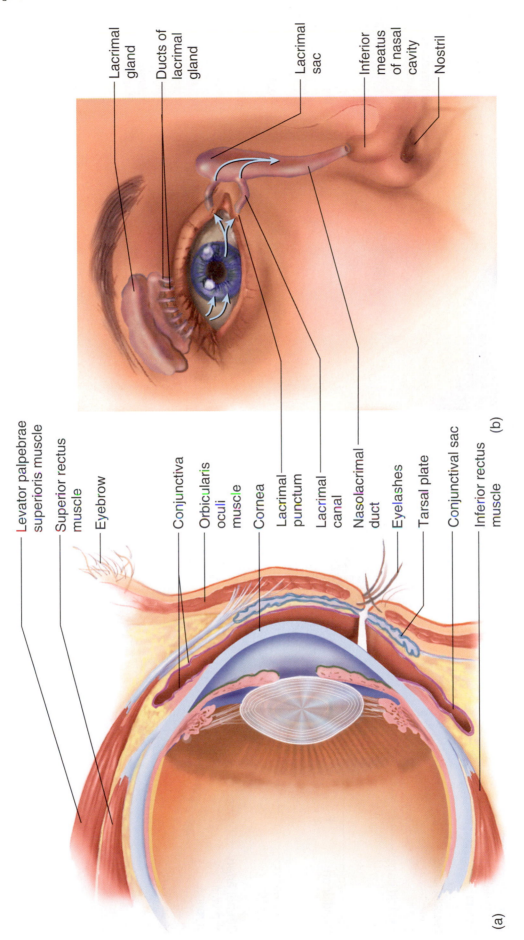

Lacrimal gland

Ducts of lacrimal gland

Lacrimal sac

Inferior meatus of nasal cavity

Nostril

Levator palpebrae superioris muscle

Superior rectus muscle

Eyebrow

Conjunctiva

Orbicularis oculi muscle

Cornea

Lacrimal punctum

Lacrimal canal

Nasolacrimal duct

Eyelashes

Tarsal plate

Conjunctival sac

Inferior rectus muscle

(a)

(b)

Figure 16.21   Accessory Structures of the Orbit

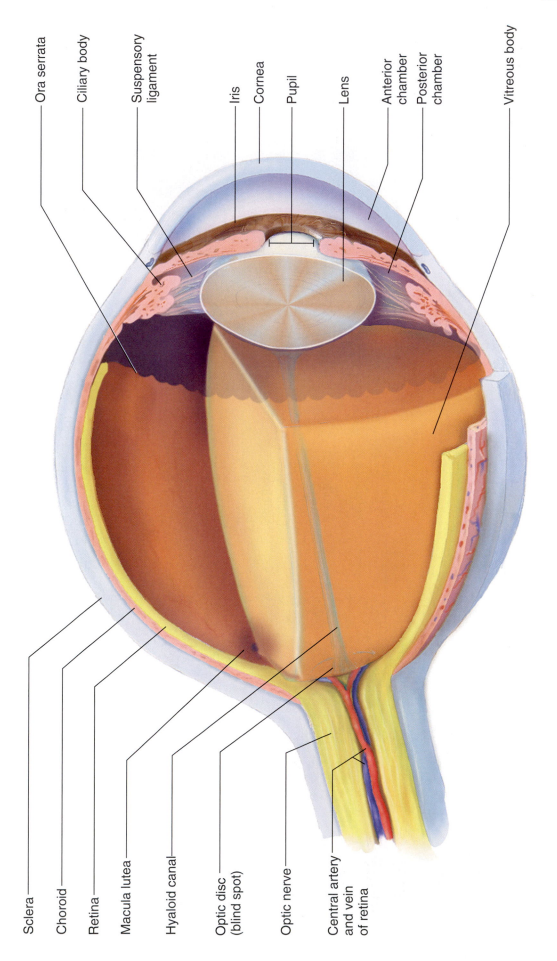

Ora serrata

Ciliary body

Suspensory ligament

Iris

Cornea

Pupil

Lens

Anterior chamber

Posterior chamber

Vitreous body

Sclera

Choroid

Retina

Macula lutea

Hyaloid canal

Optic disc (blind spot)

Optic nerve

Central artery and vein of retina

Figure 16.23 The Eye, Sagittal Section

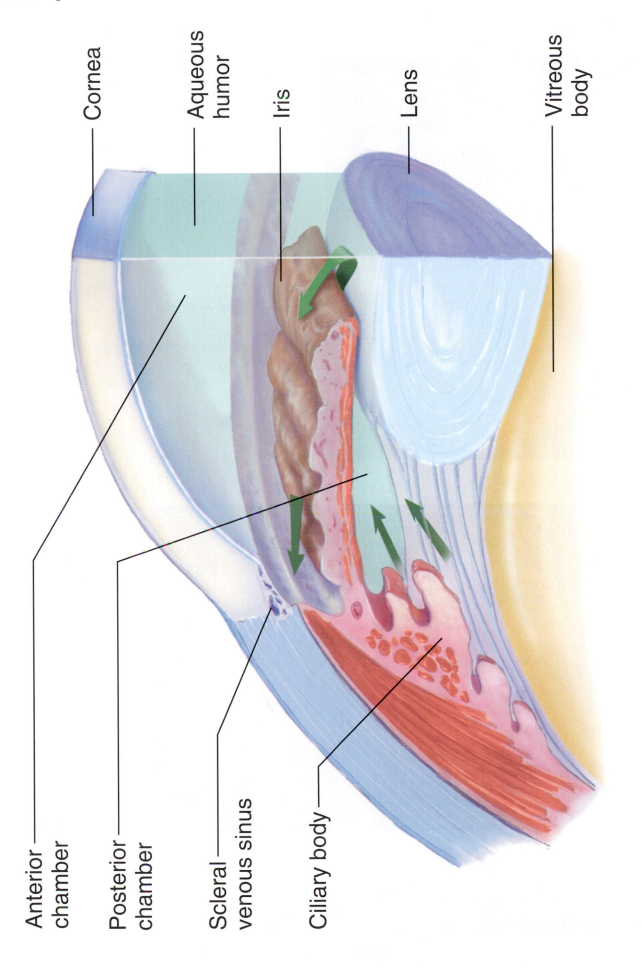

Cornea

Aqueous humor

Iris

Lens

Vitreous body

Anterior chamber

Posterior chamber

Scleral venous sinus

Ciliary body

Figure 16.24 Aqueous Humor

Unit 8: The Heart

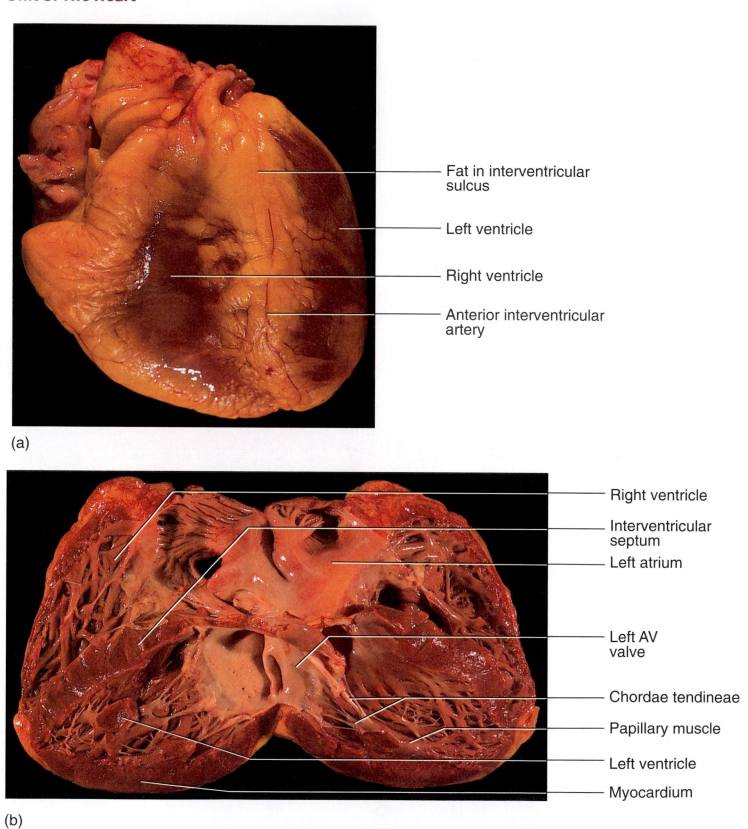

Fat in interventricular sulcus

Left ventricle

Right ventricle

Anterior interventricular artery

(a)

Right ventricle

Interventricular septum

Left atrium

Left AV valve

Chordae tendineae

Papillary muscle

Left ventricle

Myocardium

(b)

Figure 19.3   The Human Heart

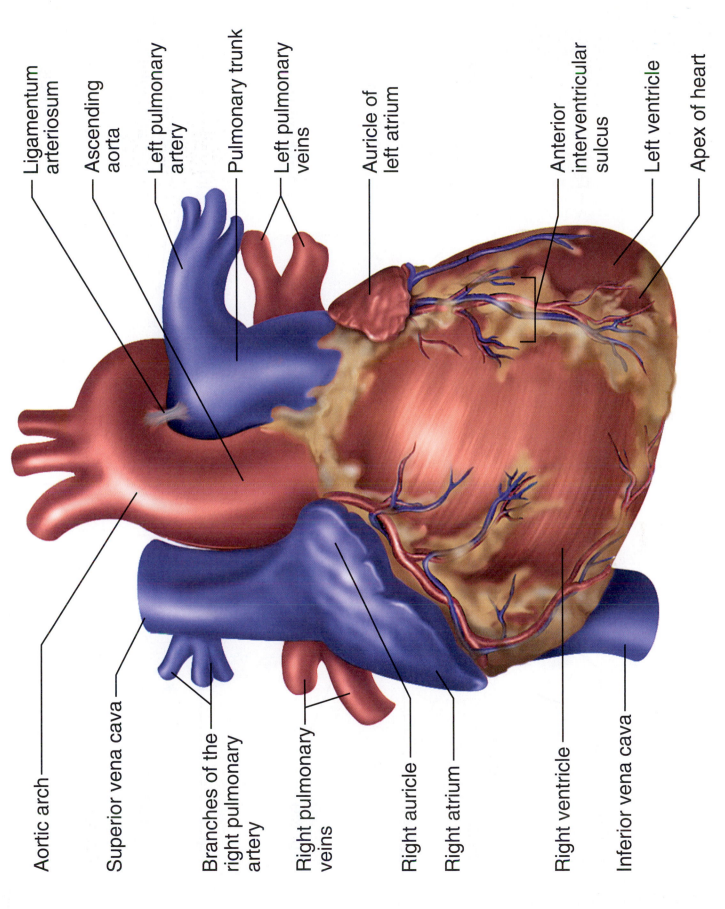

Ligamentum arteriosum

Ascending aorta

Left pulmonary artery

Pulmonary trunk

Left pulmonary veins

Auricle of left atrium

Anterior interventricular sulcus

Left ventricle

Apex of heart

Aortic arch

Superior vena cava

Branches of the right pulmonary artery

Right pulmonary veins

Right auricle

Right atrium

Right ventricle

Inferior vena cava

Figure 19.4a   External Anatomy of the Heart—Anterior View

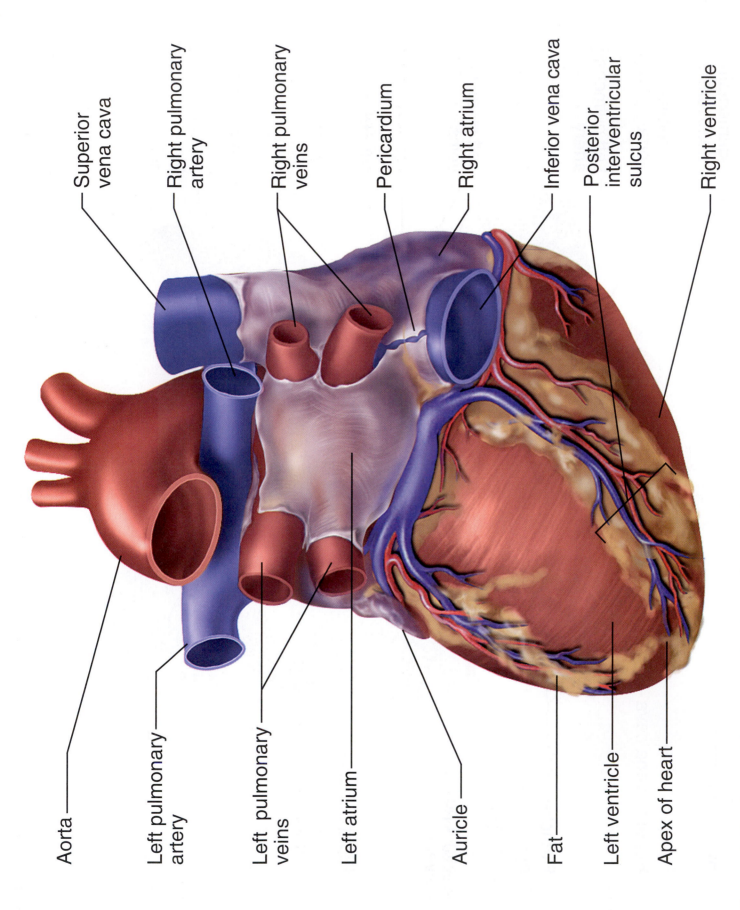

Aorta

Superior vena cava

Right pulmonary artery

Left pulmonary artery

Right pulmonary veins

Left pulmonary veins

Pericardium

Left atrium

Right atrium

Auricle

Inferior vena cava

Posterior interventricular sulcus

Fat

Left ventricle

Apex of heart

Right ventricle

Figure 19.4b   External Anatomy of the Heart—Posterior View

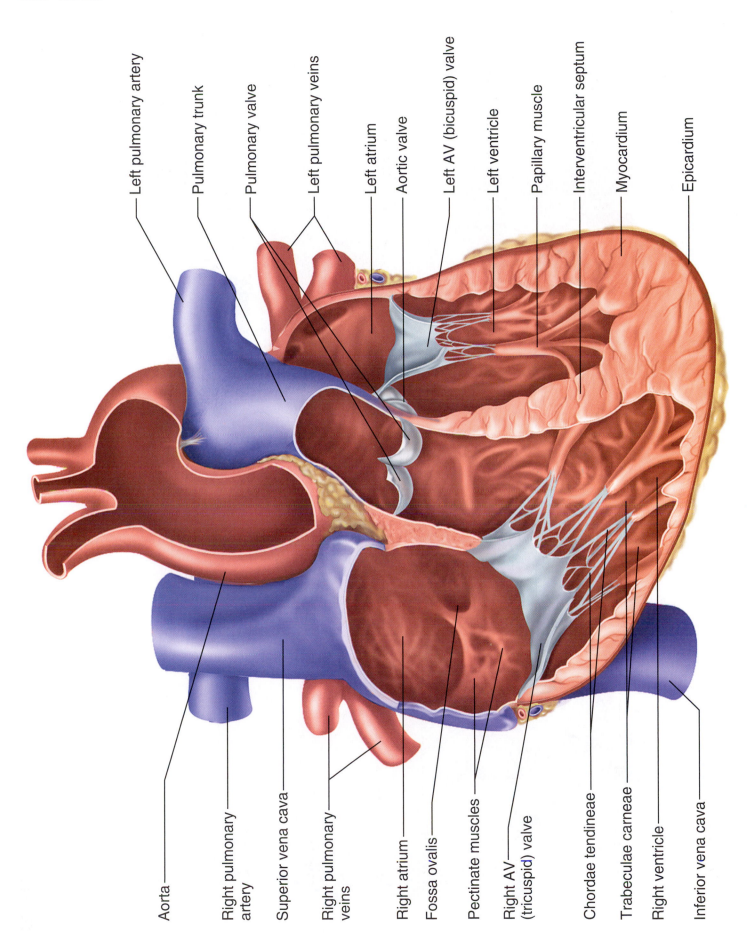

Left pulmonary artery

Pulmonary trunk

Pulmonary valve

Left pulmonary veins

Left atrium

Aortic valve

Left AV (bicuspid) valve

Left ventricle

Papillary muscle

Interventricular septum

Myocardium

Epicardium

Aorta

Right pulmonary artery

Superior vena cava

Right pulmonary veins

Right atrium

Fossa ovalis

Pectinate muscles

Right AV (tricuspid) valve

Chordae tendineae

Trabeculae carneae

Right ventricle

Inferior vena cava

Figure 19.6 Internal Anatomy of the Heart (anterior aspect)

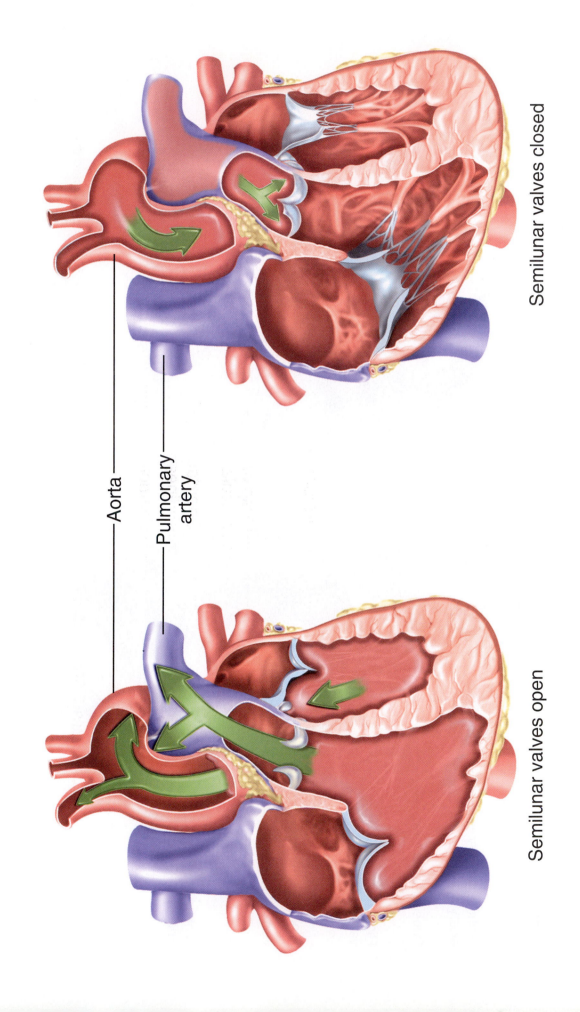

Semilunar valves closed

Aorta

Pulmonary artery

Semilunar valves open

Figure 19.8a Operation of the Heart Valves—The Semilunar Valves

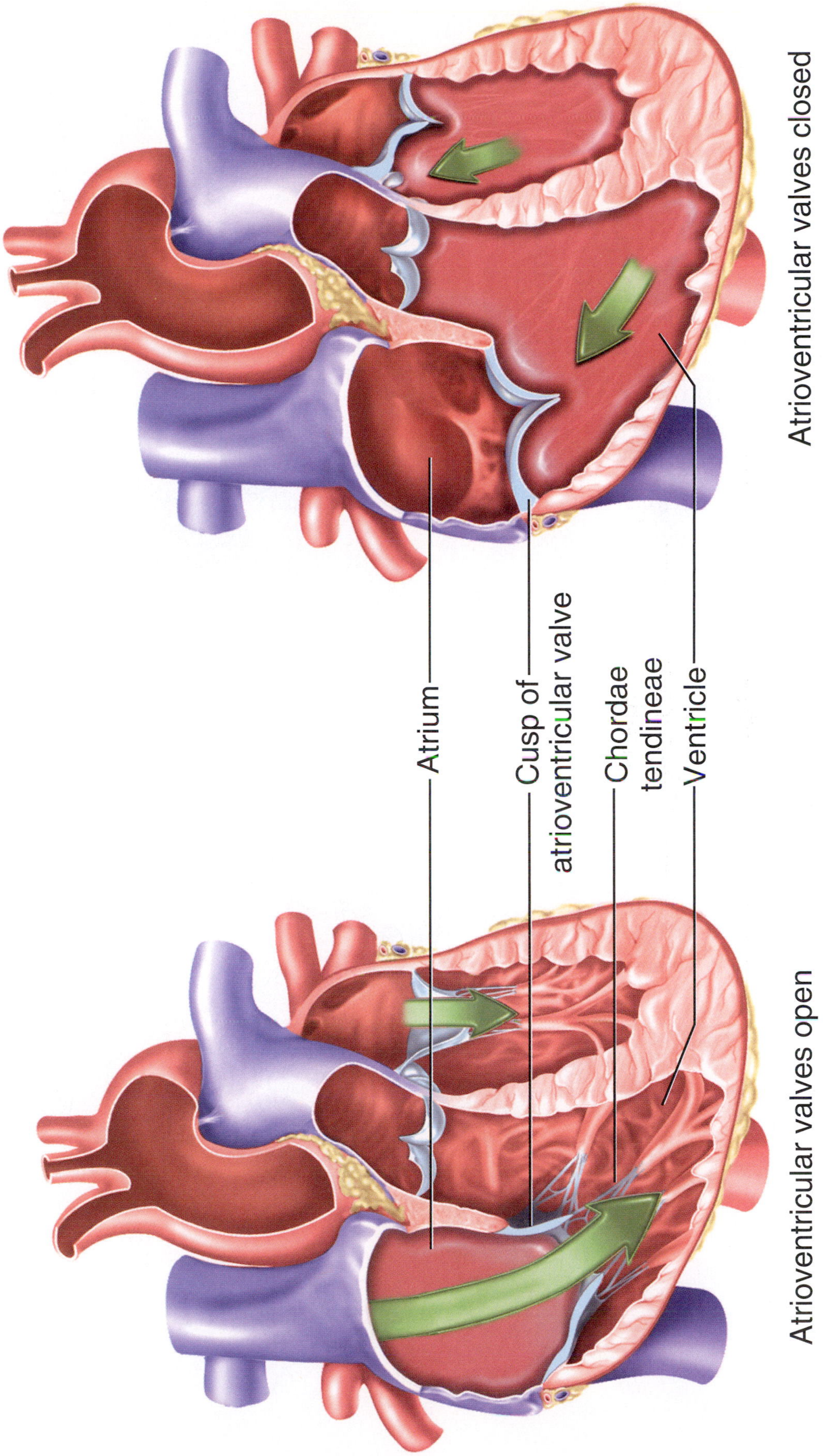

Atrioventricular valves closed

Atrioventricular valves open

Atrium

Cusp of
atrioventricular valve

Chordae
tendineae

Ventricle

Figure 19.8b   Operation of the Heart Valves—The Atrioventricular Valves

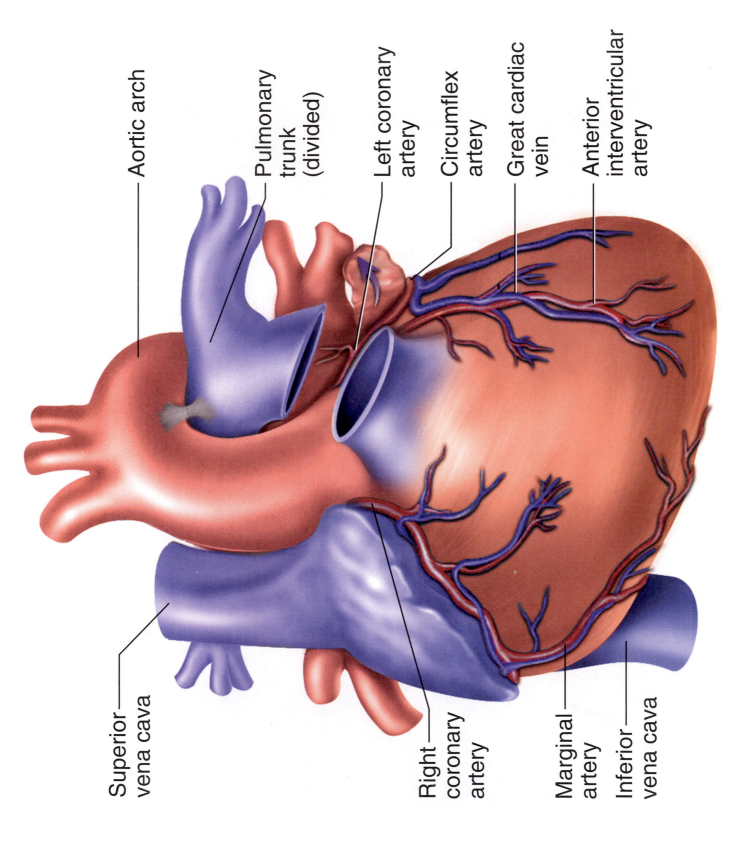

Aortic arch

Pulmonary trunk (divided)

Left coronary artery

Circumflex artery

Great cardiac vein

Anterior interventricular artery

Superior vena cava

Right coronary artery

Marginal artery

Inferior vena cava

Figure 19.10a   The Coronary Blood Vessels—Anterior View

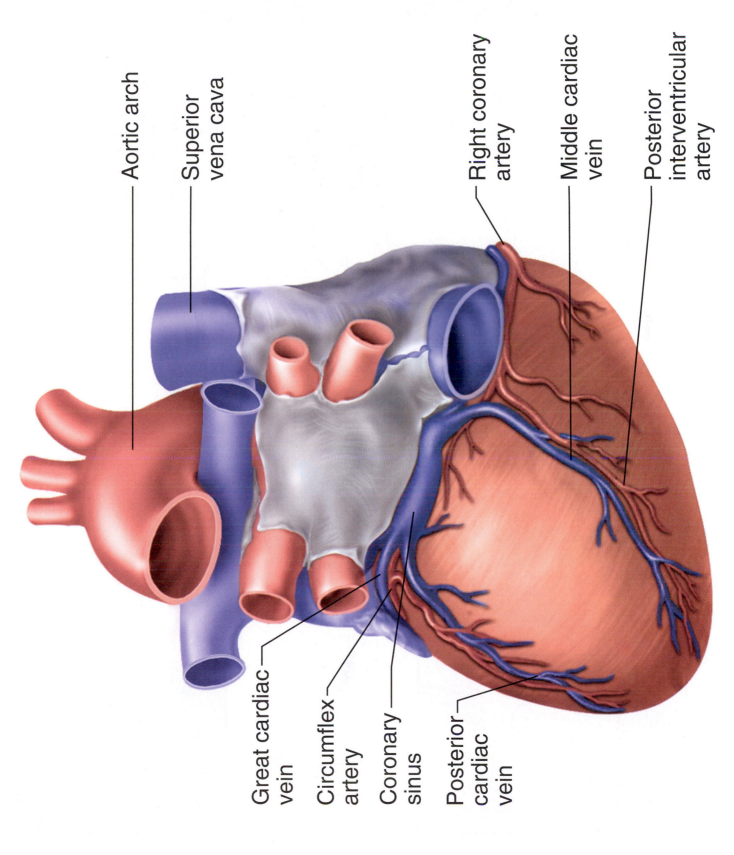

Aortic arch

Superior vena cava

Right coronary artery

Middle cardiac vein

Posterior interventricular artery

Great cardiac vein

Circumflex artery

Coronary sinus

Posterior cardiac vein

Figure 19.10b The Coronary Blood Vessels—Posterior View

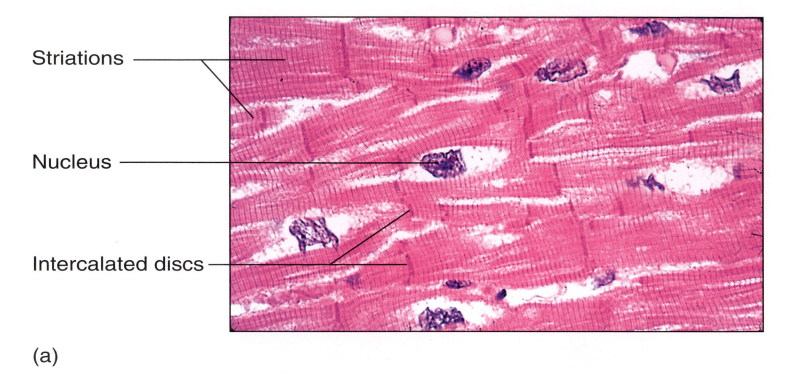

Striations

Nucleus

Intercalated discs

(a)

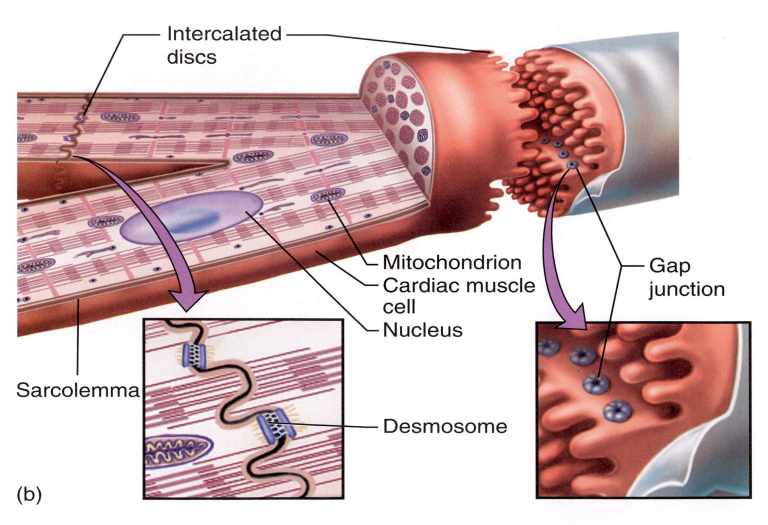

Intercalated discs

Mitochondrion

Cardiac muscle cell

Nucleus

Gap junction

Sarcolemma

Desmosome

(b)

Figure 19.11   Cardiac Muscle

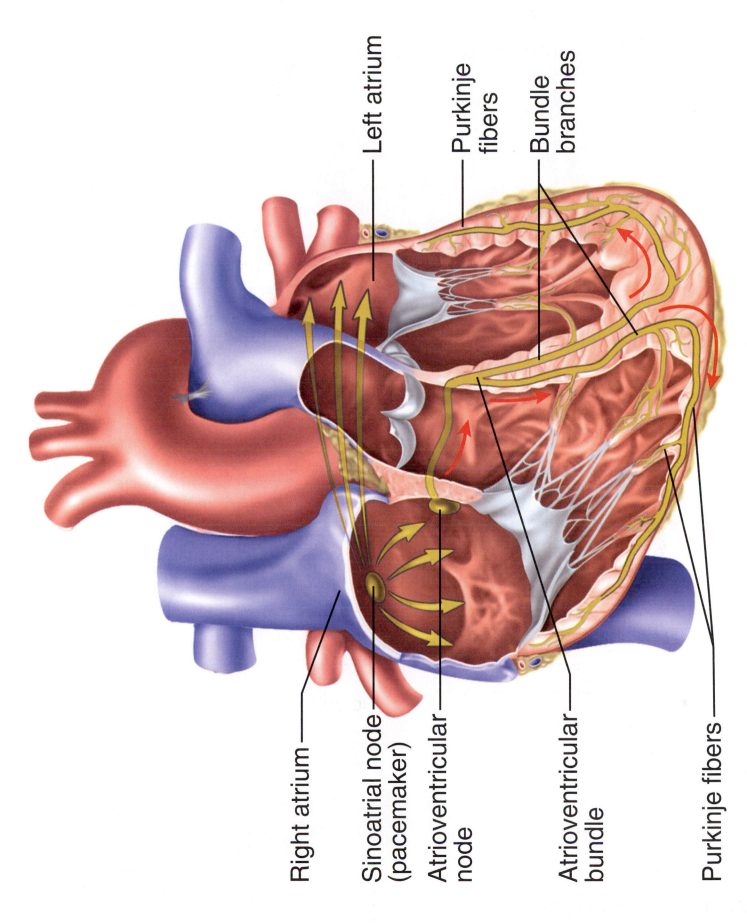

Left atrium

Purkinje fibers

Bundle branches

Right atrium

Sinoatrial node (pacemaker)

Atrioventricular node

Atrioventricular bundle

Purkinje fibers

Figure 19.12 The Cardiac Conduction System

Unit 9: The Respiratory System

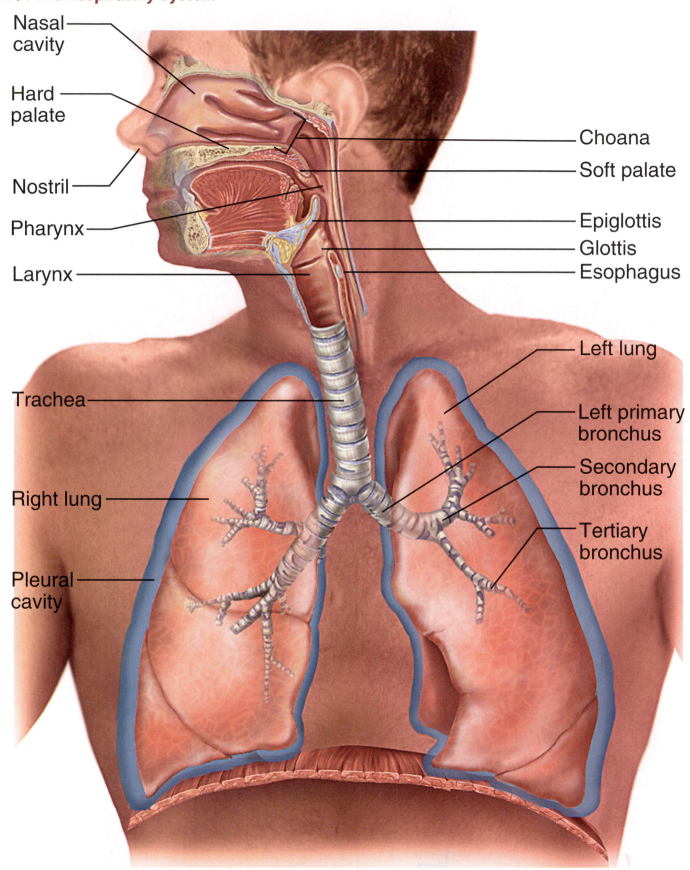

Nasal cavity

Hard palate

Nostril

Pharynx

Larynx

Choana

Soft palate

Epiglottis

Glottis

Esophagus

Trachea

Right lung

Pleural cavity

Left lung

Left primary bronchus

Secondary bronchus

Tertiary bronchus

Figure 22.1 The Respiratory System

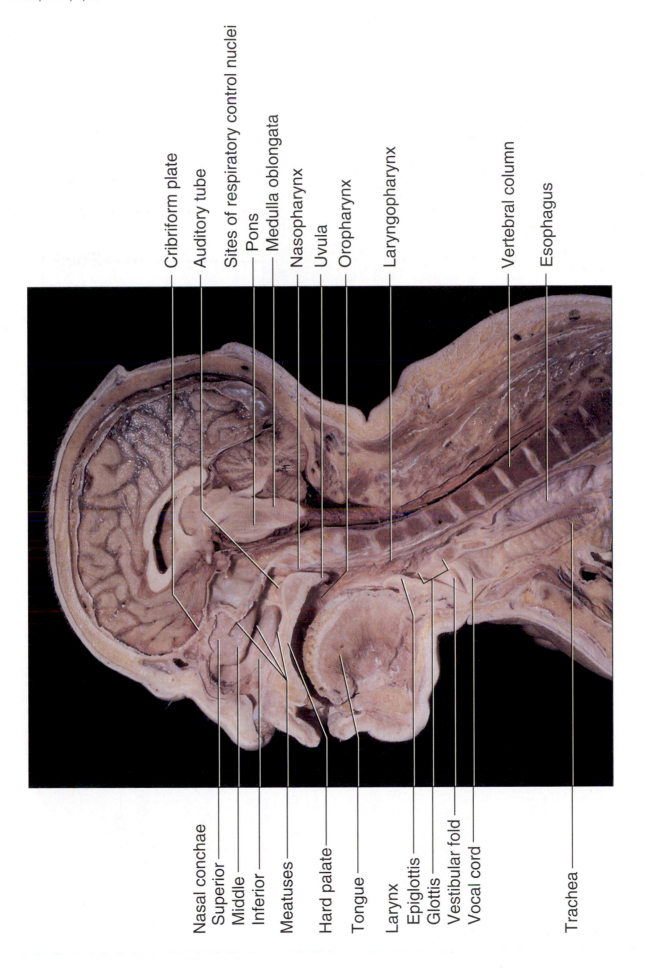

Cribriform plate
Auditory tube
Sites of respiratory control nuclei
Pons
Medulla oblongata
Nasopharynx
Uvula
Oropharynx
Laryngopharynx
Vertebral column
Esophagus

Nasal conchae
Superior
Middle
Inferior
Meatuses
Hard palate
Tongue
Larynx
Epiglottis
Glottis
Vestibular fold
Vocal cord
Trachea

Figure 22.3a Anatomy of the Upper Respiratory Tract—Median Section

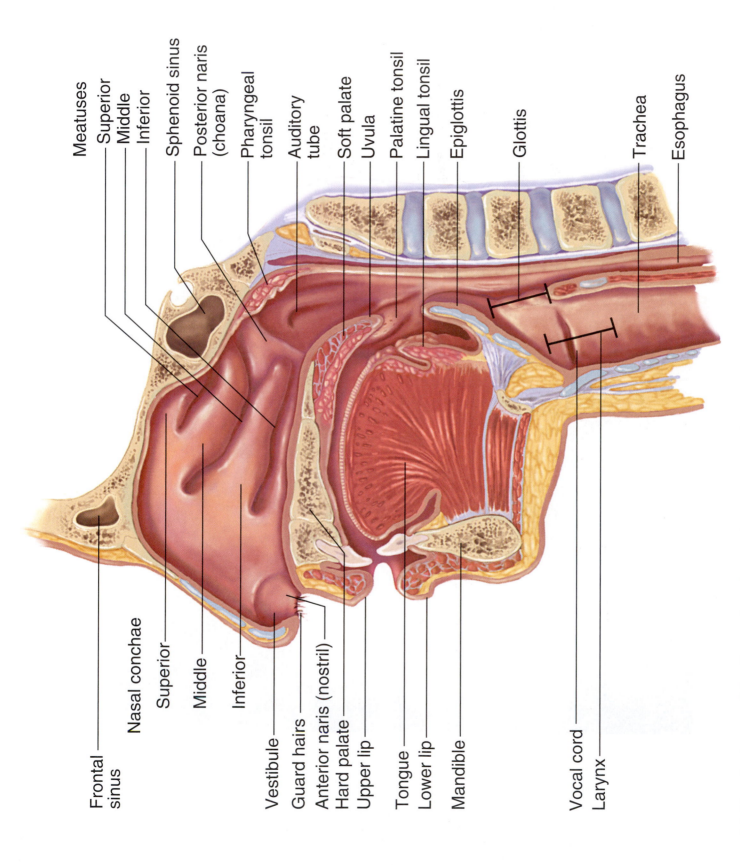

Meatuses
Superior
Middle
Inferior
Sphenoid sinus
Posterior naris (choana)
Pharyngeal tonsil
Auditory tube
Soft palate
Uvula
Palatine tonsil
Lingual tonsil
Epiglottis
Glottis
Trachea
Esophagus

Frontal sinus
Nasal conchae
Superior
Middle
Inferior
Vestibule
Guard hairs
Anterior naris (nostril)
Hard palate
Upper lip
Tongue
Lower lip
Mandible
Vocal cord
Larynx

Figure 22.3b   Anatomy of the Upper Respiratory Tract—Internal Anatomy

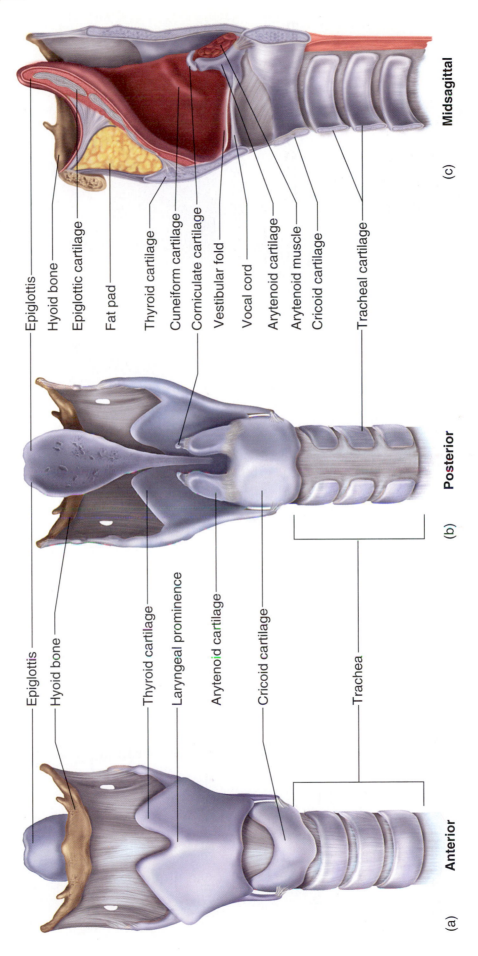

Epiglottis

Hyoid bone

Epiglottic cartilage

Fat pad

Thyroid cartilage

Cuneiform cartilage

Corniculate cartilage

Vestibular fold

Vocal cord

Arytenoid cartilage

Arytenoid muscle

Cricoid cartilage

Tracheal cartilage

Midsagittal

(c)

Epiglottis

Hyoid bone

Thyroid cartilage

Laryngeal prominence

Arytenoid cartilage

Cricoid cartilage

Trachea

Posterior

(b)

Epiglottis

Hyoid bone

Thyroid cartilage

Laryngeal prominence

Arytenoid cartilage

Cricoid cartilage

Trachea

Anterior

(a)

Figure 22.4   Anatomy of the Larynx

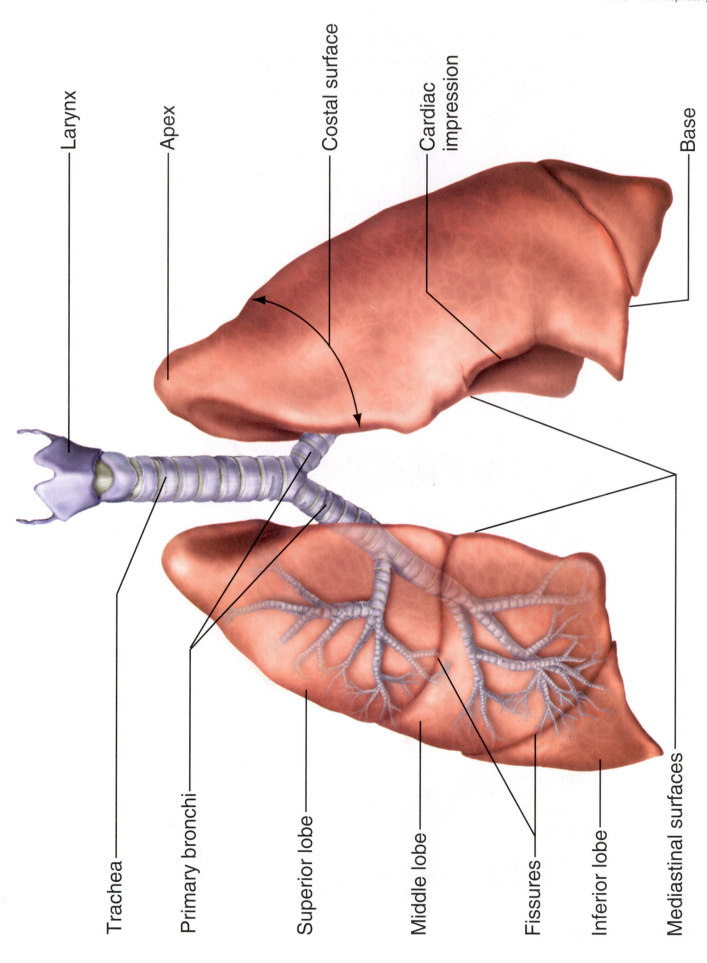

Larynx

Apex

Costal surface

Cardiac impression

Base

Trachea

Primary bronchi

Superior lobe

Middle lobe

Fissures

Inferior lobe

Mediastinal surfaces

Figure 22.9a   Gross Anatomy of the Lungs

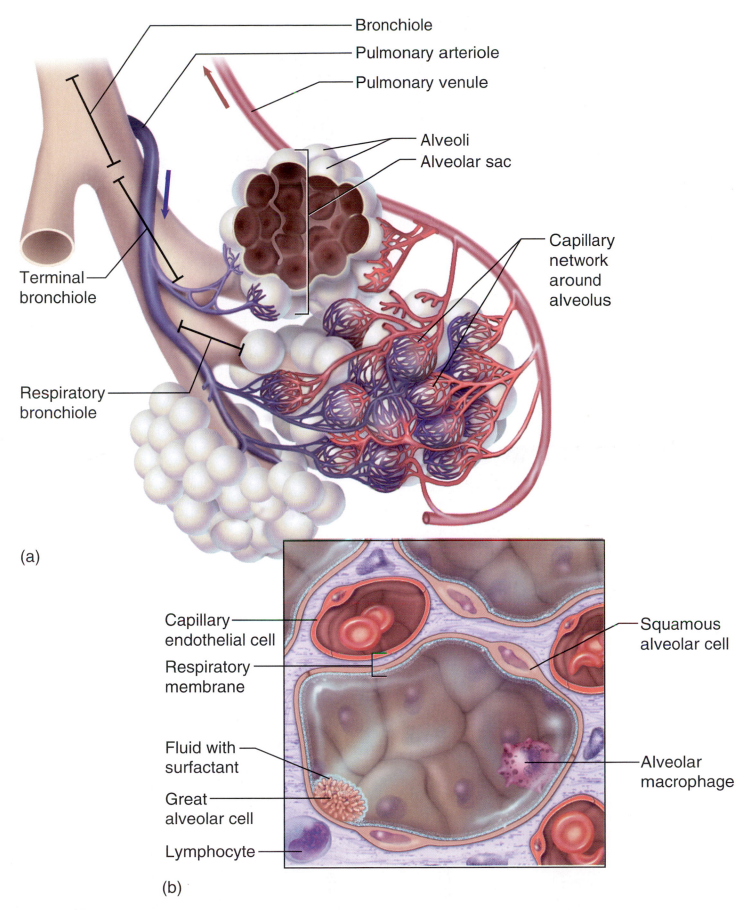

Bronchiole
Pulmonary arteriole
Pulmonary venule
Alveoli
Alveolar sac
Capillary network around alveolus
Terminal bronchiole
Respiratory bronchiole

(a)

Capillary endothelial cell
Respiratory membrane
Fluid with surfactant
Great alveolar cell
Lymphocyte
Squamous alveolar cell
Alveolar macrophage

(b)

Figure 22.12   Pulmonary Alveoli